NITRATES AND NITRITES IN FOOD AND WATER

Editor
MICHAEL HILL D.Sc., F.R.C.Path., F.R.S.C., C.Chem.
European Cancer Prevention Organisation, Hampshire, UK

WOODHEAD PUBLISHING LIMITED
Cambridge England

Published by Woodhead Publishing Limited
Abington Hall, Abington
Cambridge CB1 6AH, England

First published 1991 Ellis Horwood Limited
Reprinted 1996 Woodhead Publishing Limited

British Library Cataloguing in Publication Data
A catalogue record for this book is available from the British Library.

ISBN 1 85573 282 3

Table of Contents

Introduction

This is the age of ecology when we have at last realized that the state of our environment matters and that maintaining a healthy environment carries a price — often a very high one. We are beginning to assess that price and to brace ourselves to meet it. But before we can make the cost–benefit assessment we must first decide 'what is a healthy environment'?

I know of no example of 'environmental pollution' that does not have some positive aspects, but there is usually a strong consensus in favour of avoiding extremes. Thus, when the UK Clean Air Act removed visible air pollution in the UK during the 1960s and 1970s, gardeners and horticulturalists noted a massive increase in the level of plant disease (both fungal and aphid) and this is said to have led to a big increase in the need to use pesticides and fungicides in urban gardens. Nevertheless, few would dispute the majority view that the Clean Air Act was a very good statute for environmental protection. Similarly, only interested industrialists dispute the need to decrease or control the levels of industrial pollution of river water.

With regard to nitrate and nitrite, the perceived hazards are to the ecological balance in rivers and lakes, and to human health. Increased nitrate levels in river water lead to increased growth of algae and consequent decrease in the level of biologically available oxygen (BAO). In extreme forms, the algae form unsightly blooms on the water surface, and the BAO level falls below that necessary to support fish and other animal life. At this extreme, there is little dispute that efforts should be made to restore the ecological balance, and this is best achieved by reversing the increase in nitrate concentration that caused the problem. The health hazards are less clear but include the risk of methaemoglobinaemia in young infants and the possible risk of gastric cancer, particularly in certain high-risk patient groups.

These health risks have led to legislation to control the exposure of human to nitrate from drinking water and as a food additive. This has led to an increase in interest in the normal level of exposure to nitrate in western populations and to the realization that the nitrate content of river waters in many European countries is steadily increasing. There is considerable interest in why this should be happening, how long it has been happening for, and how can it be controlled. Where in fact, does the nitrate come from? Similarly, as horticultural methods have improved it has become possible to grow lettuce commercially under glass throughout the year. The problem is that the winter-grown lettuce contains very high nitrate levels. What can be done about this and how can the problem be solved (if, indeed, it *is* a problem)? It is normal practice to *add* nitrate and nitrite to food as a cosmetic and as a preservative. Is this necessary? If so, how much do we need to add? Are there alternatives? What effect does added nitrate have on total nitrate exposure of humans?

In this book we have attempted to answer, or at least to give the background to, some of the questions. Chapter 1 reviews the methods used to assay nitrate and nitrite. Chapters 2, 3 and 4 cover origins and the factors determining the levels of nitrate and nitrite in water. Chapter 5 is concerned with nitrate and nitrite naturally present in food, whilst Chapter 6 gives the rationale for the use of nitrate/nitrite as a food additive and its mode of action. In Chapter 7 the information on the level of human exposure to nitrate and nitrite is reviewed, together with their pharmacology and their fate in the human body. The final chapter reviews the medical evidence relating exposure to nitrate or nitrite to human disease; this covers both the acute effects of large doses and the long-term effect of chronic exposure.

On the basis of this material presented here it should be possible for those interested in the water or food industry to understand better the causes of the concerns of environmentalists, and for the environmentalist to understand better the problems associated with the control of nitrate exposure, not only with respect to decreased food production but in terms of possible wholesale changes in the appearance of the countryside.

Contributors

R. H. Dainty, PhD
MATFORSK
Norwegian Food Research Institute
Osloveien 1
N-1430 As
Norway

M. J. Hill, DSc FRCPath FRSC CChem
European Cancer Prevention Organisation
PO Box 1199
Andover
Hants SP10 1YN
UK

C. W. Keevil, PhD
Division of Pathology
PHLS-CAMR
Porton Down
Salisbury
Wiltshire SP4 OJG
UK

S. A. Leach, PhD
Division of Pathology
PHLS-CAMR
Porton Down
Salisbury
Wiltshire SP4 OJG
UK

C. Mackerness, B.Sc
Division of Pathology
PHLS-CAMR
Porton Down
Salisbury
Wiltshire SP4 OJG
UK

R. C. Massey, PhD FRSC CChem
MAFF Food Sciences Laboratory
Halden House
Queen Street
Norwich NR2 4SX
UK

D. Oakes, PhD
Water Research Centre
Medmenham
Marlow
Bucks
UK

P. J. Packer, PhD
Division of Pathology
PHLS-CAMR
Porton Down
Salisbury
Wiltshire SP4 OJG
UK

T. A. Roberts, PhD
AFRC Institute of Food Research — Reading Laboratory
Shinfield
Reading
UK

C. L. Walters, DSc FRSC CChem
Department of Biochemistry
University of Surrey
Guildford
Surrey
GU2 5XM
UK

1

Methods for the analysis of nitrate and nitrite in food and water

R. C. Massey

1.1 INTRODUCTION

The widespread occurrence of nitrate in the majority of foodstuffs and water supplies is well established (Ministry of Agriculture, Fisheries and Food 1987, Department of the Environment 1986). Concentrations vary enormously between different food items and whilst levels (wet weight) in milk are around 1 mg/kg the amounts in certain vegetables such as beetroot, celery and lettuce often exceed 1000 mg/kg. Levels of the nitrate ion can also differ markedly for different samples of the same commodity, and for instance Pickston *et al.* (1980) have reported values ranging from 630 mg/kg to 6800 mg/kg for beetroot. Similarly, nitrate concentrations may vary by over an order of magnitude in different water supplies (Department of the Environment 1986). In contrast, the presence of detectable amounts of nitrite in food and water is much rarer, and its occurrence is generally the result of microbial reduction of nitrate (National Academy of Sciences 1981, Ministry of Agriculture, Fisheries and Food 1987). It should be noted incidentally that a number of different ways have been used in the literature to express nitrate, and nitrite, concentrations including for instance mg(nitrate ion)/kg, mg(nitrate–nitrogen)/kg and mg(sodium nitrate)/kg. In addition such values may be expressed on a dry-weight or fresh-weight basis.

There are, with two exceptions, no specific regulations in the UK governing the amounts of nitrate and nitrite in foodstuffs. Nitrite, in some instances in combination with nitrate, is added to cured meats such as bacon to prevent food poisoning from *C. botulinum* and also to confer the characteristic colour and organoleptic properties to the product. The amounts that may be added are controlled by legislation (The Preservatives in Food Regulations 1989). In addition The Preservatives in Food (Amendment) Regulations 1982, Statutory Instrument (1982) No. 15, prohibit the addition of nitrate and nitrite in foods specially prepared for babies and young children.

The Water Supply (Water Quality) Regulations 1989 define the wholesomeness of drinking water supplies. These incorporate the relevant requirements of the EC

Drinking Water Directive (80/778/EEC) including the standards of 50 mg NO_3/l and for nitrite of 0.1 mg NO_2/l. These regulations cover England and Wales whilst parallel regulations operate in Scotland; equivalent regulations have not yet been made for Northern Ireland. These limits were introduced to prevent the risk of infantile methaemoglobinaemia to bottle-fed infants. Concern has also been expressed that nitrate may be a causative factor in the aetiology of gastric cancer (Correa *et al.* 1975). It is well established that a small proportion of ingested nitrate may be bacterially reduced to nitrite in the body, with the resulting formation of potentially carcinogenic N-nitroso compounds in the stomach (Ohshima and Bartsch 1981, Walters *et al.* 1982, Bonney *et al.* 1987), the large intestine (Massey *et al.* 1988, Rowland *et al.* 1991) and the infected bladder (Hicks 1982). However, there is as yet little evidence that the formation of low concentrations of these compounds in the body poses a significant threat to health, and the results of recent epidemiological studies have been equivocal (Juhasz *et al.* 1980, Armijo *et al.* 1981, Beresford 1985). It is important that the findings of such studies are treated with some circumspection, as they are potentially susceptible to the adverse influence of confounding factors including socio-economic effects and the analytical quality of the nitrate intake data.

In this chapter the parameters that govern the accuracy of nitrate and nitrite measurement in foodstuffs and water are reviewed and the currently employed analytical methods discussed.

1.2 GENERAL ANALYTICAL CONSIDERATIONS

All analytical methods are potentially susceptible to bias and the value obtained may either be higher or lower than the true result. There are two types of analytical bias: random bias and systematic bias.

1.2.1 Random bias

Random bias reflects the precision of the analytical method. If the values of repeat determinations are plotted against the number of observations then the apex of the resulting distribution curve will coincide with the true value, in the absence of systematic bias. The extent of the influence of random bias on analytical accuracy may be assessed by repeating the analysis a number of times and quantifying the standard deviation of the method. The adverse effects of this form of bias may be minimized by repeat analysis of the sample and calculation of the mean. Each stage of an analytical procedure is subject to random bias and in general the more complex and multi-step a method is the greater will be the overall standard deviation.

1.2.2 Systematic bias

In the case of systematic bias the result obtained is always higher, or lower, than the true value. Systematic errors are very much more difficult to detect than random effects and as a consequence they are the major source of inaccuracy in analytical chemistry. This form of bias is the result of either positive or negative interference on the analytical method.

1.2.2.1. Positive interference

Perhaps the most common form of positive interference arises from contamination of the sample with the analyte during analysis. Determination of the trace levels of nitrate in foodstuffs and water is particularly susceptible to this problem, as the anion may well be present as a contaminant in the reagents used for the analysis and also on filter papers and glassware. It is essential that all items and reagents used for storage and analysis are checked for nitrate prior to use. In some instances it may not be possible to completely eliminate the source of contamination. A reagent blank, taken through the entire analytical procedure, should be included in each batch of samples to quantify the blank contribution. If a response is observed for the reagent blank a question arises as to whether samples which contain very low levels of nitrate are in fact significantly different from the blank. In such circumstances it is customary to define the limit of detection of the method as three times the standard deviation of the amount of nitrate detected in the blank, and the limit of quantification (determination) as ten times the standard deviation of the blank (Keith *et al.* 1983).

Positive interference may also arise at the measurement stage. Colorimetric assays are widely employed for the determination of nitrite (British Standard 1976, Association of Official Analytical Chemists (AOAC) 1980) in which the anion is reacted to form an azo dye. Unless appropriate blanks are performed these techniques are potentially prone to interference from either naturally occurring, sample-derived, coloured pigments or species causing turbidity in the spectrophotometer cell. While these effects are of minor importance at high nitrite concentrations their relative importance increases as the limit of detection is approached. As discussed in section 1.4.2, HPLC-based methods are gaining in popularity owing to their speed and reliability; however, they too are also potentially susceptible to interference as a result of elution of components of the sample matrix at the same retention time as the analyte.

Formation of the analyte during the analytical method is not generally a problem for nitrate determination although this can occur in particular circumstances such as if the sample is solubilized by alkaline digestion (Rose, private communication). In contrast, nitrite may inadvertently be produced on storage via microbial reduction of nitrate and also during analysis by decomposition of labile species such as S-nitrosothiols (Olsman and van Leeuwen 1977).

1.2.2.2 Negative interference

Negative interferences may arise from loss of the analyte either during storage or during analysis.

Nitrite is a highly reactive species when present as its conjugate acid, nitrous acid (pK_a 3.4). Under only mildly acidic conditions it may undergo a number of reactions with other matrix components including C-, S-, O- and N-nitrosation (National Academy of Sciences 1981). In contrast, nitrate is chemically stable at the pH values that occur in foodstuffs and water. However, the anion is susceptible to microbial action and significant losses may occur on sample storage if bacterial species possessing nitrate reductase activity are present.

The concentration of the analyte will also be underestimated if incomplete extraction of the sample occurs, or if physical losses due, for instance, to adsorption take place during subsequent steps. Concentration-independent, uniformly proportional bias resulting, for example, from inefficient liquid–liquid extractions may generally be corrected for by separately spiking the sample with the analyte, assessing the recovery and correcting the analytical result accordingly. Such recovery experiments are sometimes performed by spiking a reagent blank, rather than the sample, and taking it through the analytical procedure. This approach, however, has little to commend it, as recovery losses arising from possible interaction of the components of the sample with the analyte will not be covered; similarly any losses due to sample-related modification of chromatographic behaviour will also be missed.

A different situation may arise if the recovery of the analyte is concentration dependent. This may occur for instance with recovery losses due to adsorption if there are only a limited number of active sites. Under such circumstances the recovery of the analyte is likely to increase with concentration, with the result that the recovery calculated for the spiked sample may overestimate that of the sample itself. Such phenomena may be detected by spiking at different analyte concentrations and inspecting the resulting standard addition plot for curvature.

1.2.3 Analytical quality assurance

It is a fact of life that all analytical methods are potentially at risk from bias notwithstanding the cost and sophistication of the instrumentation involved. The application of effective quality assurance procedures minimizes the possibility of such bias occurring and adversely affecting the accuracy of the analytical data. Such procedures include the use of reagent blanks, spiked samples and repeat determinations. In addition the possibility of long-term drift of results, during an extended survey for instance, can be reduced by employing an in-house reference material in each analytical batch. One very effective way of checking accuracy is the use of certified reference materials. Unfortunately whilst these products are commercially available for a number of contaminants such as heavy metals and polycyclic aromatic hydrocarbons there are as yet none certified for nitrate and nitrite. In such circumstances participation in cooperative analytical studies with other laboratories is recommended to obtain an external check on accuracy.

1.3 SAMPLE PRETREATMENT

1.3.1 Experimental design and sampling

There are a number of different sampling techniques available to estimate the dietary intake of foodstuffs and their contaminants (Ministry of Agriculture, Fisheries and Food 1988). Determination of the dietary consumption of nitrate and nitrite has often been performed by means of a total-diet study in which foods are grouped into specific categories, such as root vegetables, and each category analysed separately (Peattie *et al.* 1983, Ministry of Agriculture, Fisheries and Food 1987). Foodstuffs are purchased at retail outlets and prepared and cooked, using deionized water, in a manner similar to normal domestic practice and a composite sample of each food group is prepared and analysed. The mean dietary intake of nitrate and nitrite is calculated from the concentration found in each food group, and the relative proportion (w/w) of that group in the average total amount of food ingested per person per day. In addition to providing mean dietary intake data this approach also permits the average contribution that each food group makes to the total dietary intake of nitrite and nitrate to be assessed. If information is required on the dietary intake and distribution of individual subjects, a different approach is needed. The so-called 'duplicate diet' method may be employed in which volunteers purchase and cook twice the amount of their normal free-choice diet and the duplicate portion is reserved for analysis. This technique is susceptible to compliance problems and the possibility that participants in the scheme may alter their dietary habits (Sherlock and Walter 1983). An alternative approach, in the case of measuring nitrate intake, is the measurement of 24-hour urine samples. Packer *et al.* (1990) have recently shown that on average 55% of ingested nitrate is excreted in the urine of normochlorhydrics and have proposed that the determination of 24-hour urinary nitrate levels represents a rapid and reliable means of estimating dietary intake of the anion. This technique also has the advantage that the problems of sampling and storage of foodstuffs and beverages prior to analysis are circumvented. However, achlorhydrics need to be excluded from the sample population, as they excrete significantly less ingested nitrate than normochlorhydrics.

A number of sampling-related parameters may be expected to influence the nitrate and nitrite content of the commodity under investigation, and these factors will need to be accommodated in the overall experimental design. For instance it is well established that the nitrate content of drinking water in the UK is markedly influenced by both geological factors and fertilizer usage (Department of the Environment 1986). Sampling strategies for measurement of nitrate in lakes have been examined by Hilton *et al.* (1989), who have concluded that the commonly employed procedures do not introduce bias. The nitrate content of vegetables is dependent on a number of factors including size of the crop at harvest (Greenwood, personal communication), fertilizer form and application rate (Scaife *et al.* 1986, Greenwood and Hunt 1986), and storage time (Pickston *et al.* 1980, Machackova *et al.* 1985). In fact the nitrate concentration is not uniform throughout the plant, and Greenwood and Hunt (1986) have reported that the outer leaves of cabbage plants, which are generally not eaten, contain approximately twice the nitrate concentration of inner leaves. Similarly, uneven distribution has been observed for other vegetables including lettuce (Pickston *et al.* 1980) and carrots (Schuster and Lee 1987). Cooking of vegetables by boiling in water reduces their nitrate content by up to a factor of two (Greenwood and Hunt 1986), and equivalent results have been reported for pressure cooking (Pickston *et al.* 1980). The influence of microwave cooking has yet to be examined but it seems probable that this procedure may have little effect owing to the limited potential for leaching. Finally the analyte concentration may be influenced by manufacturing procedures employed in the production of processed foods, and Schuster and Lee (1987) have, for instance, reported that the nitrate content of carrots decreased by 32% during canning.

1.3.2 Storage

Nitrate and nitrite are both potentially unstable, and all food and water samples should be analysed as soon as possible after collection. If it is necessary to store the sample prior to analysis, the effects of the storage conditions on the stability of the analyte must be established in prior trials.

The principal concern with nitrate arises from its potential for microbial reduction by bacteria either naturally present in the sample or introduced during the sampling procedure. Nitrate may also be formed under some circumstances by oxidation of nitrite (Standing Committee of Analysts 1982); this, however, is unlikely to be a significant source of error, as the levels of nitrate naturally present in most samples are substantially higher than those of nitrite.

The magnitude of the effects of bacterial nitrate reduction may be very substantial and this is illustrated by the fact that the half-life of nitrate added to faeces is 20 mins (Saul *et al.* 1981). While the levels of bacterial contamination are orders of magnitude less in foodstuffs and their raw materials, microbial reduction may still exert a surprisingly large effect as shown in Table 1.1 (Massey and Key, unpublished data).

Table 1.1 — Measured nitrate content (mg/l) of beer wort† after storage for 3 days at:

Ambient temperature	−18°C
<1.0	50

†Hot water extract of malted barley.

Fishman *et al.* (1986) have evaluated preservation techniques for water samples and concluded that chilling (4°C) is adequate for stabilizing nitrate for up to 8 days, whilst chilling plus treatment with the biocide mercuric chloride is effective for at least 16 days. Similarly, the Standing Committee of Analysts (1982) have reported that addition of phenylmercuric acetate preserves the nitrate content of raw and potable water for over 4 weeks. This practice is, however, no longer advocated by the Standing Committee of Analysts, who recommend instead that the water be analysed as soon as possible after sampling. For biological samples such as faeces and urine it is common practice to stabilize nitrate by treatment with alkali (Archer *et al.* 1981, Packer *et al.* 1990), and this approach also has the advantage of minimizing decomposition of nitrite. There appear to be no reports of systematic investigations into the preservation of nitrate during storage of foodstuffs. In such circumstances it would seem prudent to analyse samples as soon as possible after collection. If storage is unavoidable it may be appropriate to dry or freeze the sample to prevent bacterial reduction and endogenous metabolism.

The storage of samples prior to nitrite analysis represents a major problem. The chemistry of nitrite is such that under even mildly acidic conditions a significant proportion of the anion will be in equilibrium with its conjugate acid, nitrous acid. This compound is highly reactive and will undergo a series of reactions with many nucleophilic species which are likely to be present in the sample, with the net result that the analyte is rapidly destroyed. Conversely, nitrite may be formed on storage by bacterial reduction of nitrate. The nitrite so produced may be further reduced by microbial action and will certainly be subject to further chemical decomposition. Owing to its chemical reactivity, nitrite does not occur naturally at high levels, except in cured meats, where it is added as a preservative, and is generally below the limit of detection for most commodities (Ministry of Agriculture, Fisheries and Food 1987). However, this means that errors introduced by the formation of quite modest amounts of nitrite during storage may have a significant effect on whether or not the anion is detected in the sample, and the relative magnitude of its apparent concentration.

Because of the problems associated with preserving nitrite it is recommended that samples are analysed as soon as possible after collection. The chemical decomposition of nitrite can be reduced by storing aqueous samples under strongly alkaline conditions, thus preventing the formation of nitrous acid. Such treatment also inhibits bacterial nitrate reduction. This approach is not feasible for preserving solid samples such as food, and it is also not suitable for drying, which will result in thermal decomposition of the analyte. In such circumstances it may be appropriate to store the sample at freezer temperatures. However, as shown in Table 1.2, nitrite is not indefinitely stable under these conditions, and the possible errors introduced by this approach should always be investigated in preliminary studies.

Table 1.2 — Decomposition of nitrite on storage of cured mead extracts at −18°C (from Wootton *et al.* 1985)

	Hours at −18°C		
	0	24	48
Ham	173	152	125
Salami	41	38	34
Corned beef	69	37	34

1.3.3 Extraction

A number of analytical methods for nitrite and nitrate in foodstuffs employ the same extraction procedure for both anions. Other procedures have been developed for only one of these analytes, generally nitrate, and the extraction method may not be appropriate for nitrite.

Joint extraction procedures for nitrite and nitrate have been developed for cured meat (British Standard 1976, AOAC 1984a), cheese (British Standard 1985, AOAC 1984b) and dried milk and related products (British Standard 1983): In general the sample is extracted into hot water (or borax in the case of the British Standard 1976 cured-meat method) and treated with protein precipitation reagents prior to filtration and measurement (discussed later). Dennis *et al.* (1990) reported a recovery of 92% for both nitrite and nitrate in the analysis of cured meats using a method based on the British Standard 1976 procedure. Schuster and Lee (1987) employed aqueous extraction at ambient temperature in the analysis of nitrite and nitrate in carrots, and reported mean recoveries of 95% and 87% respectively when the analytes were added prior to blending in water. Hot-water extraction was used by Wootton *et al.* (1985) in the analysis of a range of vegetables, and recoveries ranging from 62 to 104% for nitrite and 82 to 116% for nitrate were reported.

In general the procedures used to physically extract nitrate and nitrite from foodstuffs prior to the measurement stage have not been considered a source of bias in themselves. However, one aspect that has caused some concern in the past is the so-called 'dilution effect' (Usher and Telling 1975). This phenomenon involves an apparent dependence of the measured analyte concentration on the amount of water/buffer used to extract the sample. While incomplete extraction may be anticipated if insufficient solvent is used, the dilution effect results in the measured analyte concentration continuing to increase even at high (water:sample) extraction ratios. The dilution effect has very recently been reinvestigated by Rose *et al.* (1990) who have concluded that it is essentially an artefact caused by background reagent/apparatus nitrate contamination and the multiplication of this error at high dilution factors. These authors have also shown that complete enzymic digestion of a tomato paste gave identical results to hot-water and reflux-extraction procedures at 10:1 extraction ratios.

Of more concern is the certainty that nitrite will react with other components of the matrix if the extraction is carried out under even mildly acidic conditions. This effect has been shown to introduce a negative bias into the AOAC (1980) method for cured meats in which the sample is treated with hot water for 2 hours. This may be prevented by modifying the AOAC procedure such that the extraction is performed under alkaline conditions (Sen *et al.* 1979, Fox *et al.* 1982). The situation is further complicated by the fact that the analyte may be present in both free and protein-bound forms. Cassens *et al.* (1977) have reported on the basis of ^{15}N studies that of the nitrite originally added to cured meats only 5 to 20% is present as the anion in retail samples. The majority of the added nitrite reacts with other components of the matrix during curing and subsequent storage, and between 5 and 15% is estimated to interact with thiol groups, with the resulting formation of S-nitroso compounds. These species may partially dissociate to release nitrite during the extraction procedure. Olsman and van Leeuwen (1977) have developed a method to measure, separately, both free and protein (thiol)-bound nitrite by selective cleavage of the latter with mercuric chloride. Response from the bound fraction is optimized by treating the extract with mercuric ions at pH 5.0 to 6.5, the lower results observed at lower and higher pH values being attributed to reaction of nitrite with other co-extractants and precipitation of mercuric hydroxide respectively. Bousset (1980) has confirmed the increase in measurable nitrite effected by mercuric chloride for uncooked cured meats. In contrast, Fox *et al.* (1982) have concluded from work on pork slurries that mercuric chloride treatment either has little effect or reduces the amount of nitrite, depending upon the colorimetric assay employed. It is not clear, however, whether the slurries employed in this work contained levels of S-nitroso compounds typical of those present in cured meats. We have employed the method of Olsman and van Leeuwen (1977) to measure the relative amounts of free and protein (thiol)-bound nitrite in cured meats, and compared the results to those obtained using the British Standard (1976) alkaline-extraction procedure (Table 1.3).

Table 1.3 — Free and protein (thiol)-bound nitrite (mg/kg) in cured meal[†]

	Method of Olsman and van Leeuwen (1977)			British Standard (1976)
	Free nitrite	Bound nitrite	Total nitrite	Nitrite
Smoked bacon	44	60	104	87
Streaky bacon	60	21	81	72
Back bacon	67	27	94	83
Corned beef	7	4	11	11
Canned ham	4	2	6	6
Garlic sausage	49	3	52	55
Pork luncheon meat	41	5	46	47

[†]Dennis and Massey (unpublished results).

It is evident that substantial amounts of protein (thiol)-bound nitrite may be present in certain classes of cured meat. In general there is reasonable agreement between the sum of the free and protein (thiol)-bound nitrite measured by the method of Olsman and van Leeuwen (1977) and the amount of nitrite detected by the British Standard (1976) procedure. For bacon the results obtained with the British Standard (1976) method are somewhat lower, and this would appear to be a consequence of the higher amounts of protein (thiol)-bound nitrite in these samples and the possibility that the bound fraction is not completely dissociated under the alkaline conditions of the assay. The data serve to illustrate that the amount of nitrite measured in cured meats can, in certain circumstances, be significantly affected by the choice of the extraction conditions employed.

1.4 ANALYTICAL METHODS

1.4.1 Colorimetry

The most widely used colorimetric assay is that based on the reaction of nitrite, sulphanilamide and N-1-naphthylethylenediamine under acidic conditions to form a red azo dye. This forms the basis for the determination of nitrite in cured meats

(British Standard 1976, AOAC 1980), cheese (British Standard 1985, AOAC 1984b), dried milks (British Standard 1983) and raw, potable, waste waters and seawater (Standing Committee of Analysts 1982).

This assay has also found wide application for the analysis of nitrate after its prior reduction to nitrite. It is necessary for the sample to be analysed also without prior reduction in order to quantify the nitrite originally present. This is then subtracted from the total nitrite (plus nitrate) found after reduction to give nitrate. The use of cadmium columns to reduce nitrate to nitrite is widely used in the analysis of cured meats (British Standard 1976), cheese (British Standard 1985, AOAC 1984b), dried milk (British Standard 1983), fruit and vegetables (Pickston *et al.* 1980), environmental water samples (Raikos *et al.* 1988) and seawater (Standing Committee of Analysts 1982). Other nitrate-reduction techniques that have been used in conjunction with this colorimetric assay include chemical treatment with copper/hydrazine for water analysis (Standing Committee of Analysts 1982) and microbial reduction with *E. coli* for analysis of urine and foodstuffs (Packer *et al.* 1990, Hamano *et al.* 1983). Reduction/colorimetry assays are potentially prone to interference from a number of sources including incomplete reduction of nitrate to nitrite and also over-reduction of nitrate, via nitrite, to lower oxides. The general susceptibility of the colorimetric determination to chemical interference has been examined by Norwitz and Keliher (1985). High concentrations of inert salts may adversely affect the accuracy of the method as a result of the Bronsted salt effect, and the buffering effects of organic acids may affect the pH-dependent absorbance of the azo dye. In principle any component of the sample that is able to compete with the sulphanilamide reagent for nitrite during the colour development reaction will cause a negative interference. The close comparability of the results obtained for the nitrite content of cured meats by colorimetry and by HPLC, where such an interference is not anticipated, suggests that this is not a widespread problem (Eggers and Cattle 1986, Dennis *et al.* 1990). Cured meats containing high levels of ascorbate are, however, known to give low results for nitrite owing to significant competition with the coupling reagent (Fox *et al.* 1984). In addition to chemical effects the colorimetric assay may also be adversely affected by physical interferences arising from turbidity in the measurement solution. This is likely to occur if the post-extraction precipitation and filtration steps are ineffective. In the case of cured meats, Fox *et al.* (1982) recommend charcoal or alkaline extraction using a modification of the AOAC (1980) method. Detection limits for nitrate and nitrite in cured meats are generally around 1mg/kg (Fiddler *et al.* 1984, Wootton *et al.* 1985, Dennis *et al.* 1990). However, for samples that contain appreciable quantities of ascorbate or other interferants the effective limit of detection may be an order of magnitude higher (Usher and Telling 1975).

Nitrate and nitrite may also be determined by a number of other colorimetric assays. The AOAC (1984a) method for cured meats involves oxidation of nitrite to nitrate with permanganate followed by acidification and treatment with *m*-xylenol. After nitration is complete the nitroxylenol is removed from the sample by distillation and measured colorimetrically. The reaction of sulphosalicylic acid with nitrate under acidic conditions has been employed for the determination of nitrate in water but may be susceptible to interference from a number of ions including chloride

(Standing Committee of Analysts 1982). Heanes (1975) has described a method for nitrate in water based on reduction with zinc powder and colorimetric determination of the nitrite so formed by reaction with sulphanilic acid and 1-naphthol at acidic pH. More recently, Velghe and Claeys (1985) have reported a method for nitrate in water, based on its colorimetric reaction with resorcinol under acidic conditions, which shows low interference from common inorganic ions.

1.4.2 High-performance liquid chromatography

The use of high-performance liquid chromatography (HPLC) for the analysis of nitrate and nitrite has increased substantially over recent years. This is no doubt a reflection of a number of factors including the speed of analysis, the fact that both anions can be determined at the same time, toxic reagents such as cadmium are not required and also that in some instances a higher level of accuracy is obtainable as compared with older techniques.

HPLC/UV methods have been reported for the measurement of nitrate and nitrite in cured meats (Jackson *et al.* 1984, de Kleijn and Hoven 1984, Wootton *et al.* 1985, Eggers and Cattle 1986, Schreiner *et al.* 1988, Dennis *et al.* 1990), vegetables (Wootton *et al.* 1985, Hunt and Seymour 1985, Pentchuk *et al.* 1986, Schuster and Lee 1987), beer (Jancar *et al.* 1984, Massey *et al.* 1990) and water (Iskandarani and Pletrzyk 1982, Kok *et al.* 1983, Hern *et al.* 1983, Schroeder 1987, Mullins 1987, Frohlich 1987). The majority of workers have employed direct UV detection of the column eluant to determine nitrate and nitrite. However, for multi-anion measurement this approach is not suitable owing to the poor UV absorbance of species such as chloride and phosphate. In such cases the analytes may be measured by indirect UV in which a UV-absorbing compound is included in the mobile phase and anions having a lower absorbance give rise to a negative peak on elution from the column. This technique has been used for multi-anion analysis of water by a number of investigators (Frohlich 1987, Mullins 1987). An alternative approach involves the application of a conductivity, detector, and this has been employed in the analysis of vegetables (Pentachuk *et al.* 1986), beer (Jancar *et al.* 1984) and water (Hern *et al.* 1983). Ion-exchange HPLC methods for the multi-anion analysis of water using a number of detectors, including direct/indirect UV and direct and suppressed conductivity, have also very recently been published by the Standing Committee of Analysts (1990).

Reverse-phase (ion interaction/ion pair) HPLC has been used in the analysis of water (Frohlich 1987, Mullins 1987, Schroeder 1987). However, ion-exchange HPLC is preferred by the majority of workers for foodstuff analysis (Hunt and Seymour 1985, Pentachuk *et al.* 1986, Dennis *et al.* 1990). Jackson *et al.* (1984) have compared the applicability of both reverse-phase and ion-exchange HPLC to the analysis of foods. They concluded that the former technique is not sufficiently robust, being prone to repeatability problems and showing variation between different sample types. The maintenance of chromatographic performance is a general problem in the analysis of complex food extracts, as components of the matrix may adsorb onto the column and adversely affect subsequent analyses. Dennis *et al.* (1990) have used disposable cyclohexyl Bond Elut cartridges in the analysis of cured meats to clean the sample extract prior to HPLC measurement. Hunt and Seymour

(1985) treated vegetable extracts with activated charcoal to adsorb organic species which might otherwise have interfered with the chromatographic analysis.

The detection limits for HPLC-based methods for nitrite and nitrate are typically in the range 0.01 to 0.1 mg/l for water (Kok *et al.* 1983, Schroeder 1987) and between 0.1 and 1.0 mg/kg for foodstuffs (Eggers and Cattle 1986, Dennis *et al.* 1990). In practice, of course, the detection limit is dependent on the precision of the reagent blank.

As part of the validation of their methods a number of workers have compared HPLC results with those obtained by other procedures. In the case of water, good agreement with colorimetric results has been reported by Schreiner *et al.* (1988), Schroeder (1987) and Iskandarani and Pletrzyk (1982). Discrepancies between HPLC and a direct UV screening method were observed by Kok *et al.* (1983) and attributed to interferences to the latter technique. Hunt and Seymour (1985) reported excellent agreement between HPLC and a Devarda's alloy method for the measurement of nitrate in a range of vegetables. A similar finding was noted by Pentachuk *et al.* (1986), who compared the analysis by HPLC and colorimetry of a number of types of vegetable. In contrast, HPLC was found to give higher results than colorimetry in the analysis of nitrate in carrots, and this was thought likely to be due to incomplete colour development (Schuster and Lee 1987). Schreiner *et al.* (1988) have reported good agreement between HPLC and colorimetry for the analysis of both nitrite and nitrate in cured meats. Similar findings were noted by Dennis *et al.* (1990) although the concordance between the two methods was poorer below 10 mg/kg. Excellent agreement of the results of these two techniques was observed by Eggers and Cattle (1986) for the nitrite content of cured meats. Agreement for nitrate was more modest, and this was thought to be due to variable blanks and incomplete recoveries associated with the cadmium-reduction step. Relatively poor comparability of HPLC and colorimetric results for nitrite in a range of cured meats was observed by Wootton *et al.* (1985). These authors employed reverse-phase HPLC, in contrast to the majority of other workers, who used ion exchange, and this mode of separation may be less suited to complex matrices such as cured meat.

1.4.3 Gas chromatography

A number of gas chromatography (GC) methods have been published for the measurement of nitrate (Wu and Saschenbrecker 1977, Tanaka *et al.* 1982, Ross and Hotchkiss 1985) and nitrite (Funazo *et al.* 1980, Tanaka *et al.* 1980, Tanaka *et al.* 1982, Wu *et al.* 1984) in water and foodstuffs. These procedures all involve the formation of a volatile derivative, extraction into organic solvent and measurement by GC using a selective detector. Wu *et al.* (1984) have developed a GC/electron-capture detector method for the pentafluorobenzyl derivative of nitrite. The procedure has a detection limit of 5 μg/l for aqueous samples, and good agreement with a colorimetric assay was reported for the analysis of cured meat, saliva and water samples. Similar performance characteristics were also achieved by Funazo *et al.* (1980) in their method for nitrite involving GC/electron-capture detection of bromochlorobenzene formed by reaction of nitrite, bromoaniline and cupric chloride. Nitration of benzene under acidic conditions followed by GC/electron-capture detection forms the basis of the method of Wu and Saschenbrecker (1977) for nitrate in meat products. Ross and Hotchkiss (1985) used the nitrobenzene derivative for the measurement of nitrate in dried foods, and reported a detection limit of 100 μg/kg.

1.4.4 Other methods

Doerr *et al.* (1981, 1982) have developed a chemiluminescence method for nitrite in foods, involving reduction of nitrite by treatment with ascorbic acid under acidic conditions, followed by chemiluminescence detection of the released nitric oxide. The detection limit of the method for dried milks was 25 μg/kg, and good agreement with a colorimetric assay was obtained. Fiddler *et al.* (1984) compared the results obtained for the nitrite content of cured meats by this method and by colorimetry. No significant differences were observed when 1-naphthylamine/sulphanilamide was used as the colour reagent although the data obtained with 1-naphthylethylenediamine/sulphanilamide were on average 25% lower. This discrepancy was attributed to likely interference by ascorbate in the sample on the colour development of the latter reagent. Walters *et al.* (1987) employed a somewhat different approach in their chemiluminescence assay for nitrate and nitrite. In this method the aqueous sample or extract is injected into sodium iodide/acetic acid for measurement of nitrite, or into ferrous ammonium sulphate/ammonium molybdate/sulphuric acid for determination of nitrate plus nitrite. While the smaller sample sizes involved with this method lead to poorer detection limits, *ca.* 150 μg/l, than the procedure of Doerr *et al.* the throughput rate is considerably higher.

Nitrate plus nitrite may be reduced with Devarda's alloy under alkaline conditions and the resulting ammonia distilled and determined by titration. The method is free from interferences, with the exception of high levels of amides which may slowly decompose to release traces of ammonia during distillation, and has a detection limit of 0.2 mg/l for water (Standing Committee of Analysts 1982). The procedure has also been used in the analysis of vegetables by Hunt and Seymour (1985) who reported good agreement with results produced by HPLC. The determination of nitrate by direct UV spectrophotometry represents a rapid method of analysis in principle. Unfortunately, however, it is subject to severe interference from organic matter and several ionic species; as a result its application is limited to non-saline waters which are low in organic matter and contain high levels of nitrate (Standing Committee of Analysts 1982). A fluorimetric method has been reported by Motomizu *et al.* (1987) for the determination of nitrate in water, which involves copperized cadmium reduction to nitrite followed by reaction with 3-amino-1,5-naphthalenedisulphonic acid to form the azoic acid. The detection limit of the procedure is 0.5 mg/l, and the method showed good agreement with HPLC. Damiani and Burini (1986) have described a fluorimetric assay for nitrite with a similar detection limit based on its reaction with 2,3-diaminonaphthalene. Application of the method to milk samples gave results that were approximately 10% lower than a colorimetric assay. Capillary isotachophoresis has been employed by Kaniansky *et al.* (1986) in the analysis of rainwater, and good agreement was shown with results obtained by colorimetry. Nitrate-selective electrodes have found little application in the analysis of foodstuffs because of their potential interference from several commonly occurring anions such as chloride, sulphate and bicarbonate. Pentachuk *et al.* (1986) have reported that nitrate electrodes give rise to significant positive interference when compared to other methods in the analysis of vegetables. For simpler matrices such as sugar (Karim *et al.* 1986) and water (Standing Committee of Analysts 1982) the use of electrodes may be possible, if relatively poor revision is acceptable and providing that the results are cross-checked against a more selective procedure. Mertens *et al.* (1975) have described the use of an ammonia probe for the determination of nitrate in water, in which the analyte is first reduced to ammonia by Devarda's alloy. Prior flushing of the sample with nitrogen under alkaline conditions is necessary to remove ammonia naturally present in the sample. The method has a detection limit to about 70 μg/l, and showed good agreement with the results of a colorimetric assay. Other methods that have been reported for the analysis of nitrate and nitrite include ion pair extraction/atomic absorption spectrophotometry (Silva *et al.* 1986), differential pulse polarography (Barsotti *et al.* 1982) and stripping voltammetry (Van den Berg and Li 1988).

1.5 CONCLUSIONS

The measurement of nitrite and, to a lesser extent nitrate, in foodstuffs and water represents an interesting challenge to the analyst. Unless careful precautions are taken, systematic bias may be introduced at all stages of the analytical procedure including sampling, storage, extraction and the end-determination step.

Colorimetric assays, particularly the reaction with sulphanilamide and *N*-1-naphthylethylenediamine, have been widely used for the measurement of nitrite, and nitrate after prior reduction to nitrite. In general the technique gives results that agree well with other procedures. However, at analyte concentrations below 20 mg/kg there are a number of instances when the concordance is poorer. This has often been attributed to problems of chemical interference or turbidity with the colorimetric method. Over recent years newer techniques such as HPLC, derivatization/GC and chemiluminescence have been developed and applied with success to foodstuffs and water. Of these procedures HPLC is the most widely used, and this is no doubt a reflection of the speed and reliability of the results produced and the availability of the necessary instrumentation in most trace-analytical laboratories.

The reliable determination of nitrite is something of a problem. With the exception of cured meats, the analyte is generally present at very low levels and potentially subject to interferences at the measurement stage. Under such circumstances it is possible to assign the observed detector response incorrectly to nitrite and introduce a positive bias. In the case of cured meats the extent to which species such as S-nitrosothiols dissociate to nitrite during the extraction stage will affect the analytical result obtained. As a consequence the amount of nitrite measured is dependent on the extraction procedure employed.

In contrast to the substantial literature on analytical methods for nitrate and nitrite, and their application to foodstuffs and water, there is a dearth of information on the bias introduced by incorrect storage of samples prior to analysis. Indeed it is the case that published surveys often detail the analytical method employed in full but provide very little information on the sample storage conditions used and the possible resulting bias. It is well established that nitrate is prone to microbial reduction and may also be produced or utilized as a result of post-sampling metabolic processes that may occur for instance in vegetables. Nitrite is chemically highly reactive and its concentration may be expected to drop rapidly on storage particularly under even mildly acidic conditions. In such circumstances it is clearly prudent to perform the analysis as soon as possible after sampling. However, there may be occasions when this is impractical such as when the sampling is done at a distance from the laboratory, possibly by staff from a different organization. In this situation the effects of the proposed storage conditions on the analyte concentration will need to be checked before conducting the survey. For foodstuffs in particular there appear to be very few reported investigations into procedures for stabilizing nitrate and nitrite during storage prior to analysis, and this deficiency needs to be remedied.

REFERENCES

Archer, M. C., Saul, R. L., Lee, L.-J. and Bruce, W. R. (1981). Analysis of nitrate, nitrite and nitrosamines in human faeces. In: *Gastrointestinal Cancer: Endogenous Factors*. Ed. W. R. Bruce. Banbury Report No. 7, Cold Spring Harbor Laboratory, pp. 321–332.

Armijo, R., Orellana, M., Mediona, E., Coulson, A. H., Sayre, J. W. and Detels, R. (1981). Epidemiology of gastric cancer in Chile: I. Case-control study. *Int. J. Epidemiology* **10**, 53–56.

Association of Official Analytical Chemists (1980). *Nitrites — Official First Action.* Section 24.041, 13th edn, AOAC, Arlington, USA.

Association of Official Analytical Chemists (1984a). *Nitrates and Nitrites in Meat, Xylenol Method, First Action*. Section 24.041, 14th edn, AOAC, Arlington, USA.

Association of Official Analytical Chemists (1984b). *Nitrate and Nitrite in Cheese, Modified Jones Reduction, Final Action.* Section 16.278, 14th edn, AOAC, Arlington, USA.

Barsotti, D. J., Pylypiw, H. M. and Harrington, G. W. (1982). The determination of nitrite and nitrate ions by differential pulse polarography. *Analyt. Lett.* **15**, 1411–1422.

Beresford, S. A. A. (1985). Is nitrate in the drinking water associated with the risk of cancer in the urban UK. *Int. J. Epidemiology* **14**, 57–63.

Bonney, G. E., Elston, R. C., Correa, P., Tannenbaum, S. R., Haenszel, W., Zavala, D. E., Fontham, E., Zarama, G., Gordillo, G. and Cuello, C. (1987). Genetic etiology of gastric cancer: II. Segregation analysis of gastric pH, nitrate and nitrite. *Genetic Epidemiology* **4**, 103–114.

Bousset, J. (1980). Die rolle des quecksilberchlorids bei der nitritbestimmung. *Fleischwirtschaft* **60**, 1325–1326.

British Standard (1976). Meat and meat products. Part 7, determination of nitrate content. Part 8, determination of nitrite content. BS 4401. The British Standards Institute, London.

British Standard (1983). Analysis of dried milk and dried milk products. Part 16, determination of nitrate and nitrite contents. Section 16.1, method for dried milk by cadmium reduction. BS 1743. The British Standards Institute, London.

British Standard (1985). Chemical analysis of cheese. Part 9, determination of nitrate and nitrite contents. Cadmium reduction and photometry method. BS 770. The British Standards Institute, London.

Cassens, R. G., Woolford, G., Lee, S. H. and Goutefongea, R. (1977). Fate of nitrite in meat. In: *Proceedings of the Second International Symposium on Nitrite in Meat Products*, Eds B. J. Tinbergen and B. Krol, pp. 95–100, Pudoc, Wageningen.

Correa, P., Haenszel, W., Cuello, C., Tannenbaum, S. and Archer, M. (1975). A model for gastric cancer epidemiology. *Lancet* **ii**, 58–60.

Damiani, P. and Burini, G. (1986). Fluorimetric determination of nitrite. *Talanta* **33**, 649–652.

de Kleijn, J. P. and Hoven, K. (1984). Determination of nitrite and nitrate in meat products by high-performance liquid chromatography. *Analyst* **109**, 527–528.

Dennis, M. J., Key, P. E., Papworth, T., Pointer, M. and Massey, R. C. (1990). The determination of nitrate and nitrite in cured meats by HPLC/UV. *Food Add. Contam.* in the press.

Department of the Environment (1986). *Nitrate in Water.* Pollution Paper No 26. HMSO, London.

Doerr, R. C., Fox, J. B., Lakritz, L. and Fiddler, W. (1981). Determination of nitrite in cured meats by chemiluminescence. *Anal. Chem.* **53**, 381–384.

Doerr, R. C., Gates, R. A. and Fiddler, W. (1982). Chemiluminescence detection of nitrite in non-fat dried milk powders. *J. Ass. Off. Anal. Chem.* **65**, 616–618.

Eek, L. and Ferrer, N. (1985). Sensitive determination of nitrite and nitrate by ion-exchange chromatography. *J. Chrom.* **322**, 491–497.

Eggers, N. J. and Cattle, D. L. (1986). High performance liquid chromatographic method for the determination of nitrate and nitrite in cured meat. *J. Chrom.* **354**, 490–494.

Fiddler, W., Doerr, R. C., Gates, R. A. and Fox, J. B. (1984). Comparison of chemiluminescent and AOAC methods for determining nitrite in commercial cured meat products. *J. Assc. Off. Anal. Chem.* **67**, 525–528.

Fishman, M., Schroder, L. J. and Shockey, M. W. (1986) Evaluation of methods for preservation of water samples for nutrient analysis. *Int. J. Ev. Studies* **26**, 231–238.

Fox, J. B., Doerr, R. C. and Lakritz, L. (1982). Interaction between sample preparation techniques and three methods of nitrite determination. *J. Assc. Off. Anal. Chem.* **65**, 690–695.

Fox, J. B., Doerr, R. C. and Gates, R. (1984). Effect of residual ascorbate on determination of nitrite in commercial cured meat products. *J. Assc. Off. Anal. Chem.* **67**, 692–697.

Frohlich, D. H. (1987). HPLC of anions on coated reversed phase columns using indirect UV-detection and eluants containing nitrophthalic acids. *J. High Res. Chromat. and Chromat. Comm.* **10**, 12–16.

Funazo, K., Tanaka, M. and Shono, T. (1980). Determination of nitrite at parts-per-billion levels by derivatization and electron capture gas chromatography. *Anal. Chem.* **52**, 1222–1224.

Greenwood, D. J. and Hunt, J. (1986). Effect of nitrogen fertiliser on the nitrate contents of field vegetables grown in Britain. *J. Sci. Food Agric.* **37**, 373–383.

Hamano, T., Mitsuhashi, Y., Tanaka, K., Matsuki, Y., Oji, Y. and Okamoto, S. (1983). Application of nitrate reductase for the determination of nitrate in meat and fishery products. *Agric. Biol. Chem.* **47**, 2427–2433.

Heanes, D. L. (1975). Determination of nitrate in soil and water by an adaptation of an Orange 1 method. *Analyst* **100**, 316–321.

Hern, J. A., Rutherford, G. K. and van Loon, G.W. (1983). Determination of chloride, nitrate, sulphate and total sulphur in environmental samples by single-column ion chromatography. *Talanta* **30**, 677–682.

Hicks, R. M. (1982). Nitrosamines as possible etiological agents in bilharzial bladder cancer. In: *Nitrosamines and Human Cancer*, Ed. P. Magee, Banbury Report 12, Cold Spring Harbour Laboratory, pp. 455–472.

Hilton, J., Carrick, T., Rigg, E. and Lishman, J. P. (1989) Sampling strategies for water quality monitoring in lakes: the effect of sampling method. *Env. Pollution* **57**, 223–224.

Hunt, J. and Seymour, D. J. (1985). Method for measuring nitrate-nitrogen in vegetables using anion-exchange high-performance liquid chromatography. *Analyst* **10**, 131–133.

Iskandarani, Z. and Pletrzyk, D. J. (1982). Determination of nitrite and nitrate in water and food samples by ion interaction chromatography. *Anal. Chem.* **54**, 2601–2603.

Jackson, P. E., Haddad, P. R. and Dilli, S. (1984). Determination of nitrate and nitrite in cured meats using high-performance liquid chromatography. *J. Chrom.* **295**, 471–478.

Jancar, J. C., Consant, M. D. and Herwig, W. C. (1984). Determination of inorganic anions in beer by ion chromatography. *J. Amer. Soc. Brewing Chem.* **42**, 90–93.

Juhasz, L., Hill, M. J. and Nagy, G. (1980). Possible relationship between nitrate in drinking water and the incidence of stomach cancer. In: *N-Nitroso Compounds: Analysis, Formation and Occurrence*. Eds E. A. Walker, M. Gastegnaro, L. Griciute and M. Borzsonyi, International Agency for Research on Cancer (Lyon), pp. 619–623.

Kaniansky, D., Zelensky, I., Havasi, P. and Cerovsky, M. (1986). Determination of nitrate and sulphate in rain-water by capillary isotachophoresis. *J. Chrom.* **367**, 274–279.

Karim, M. T., Najib, F. M. and Mohammad, M. S. (1986). Determination of chloride, nitrate and calcium ions in sugar with ion selective-electrodes. *J. Food Tech.* **21**, 559–568.

Keith, L. H., Crummett, W., Deegan, J., Libby, R. A., Taylor, J. K. and Wentler, G. (1983). Principles of environmental analysis. *Anal. Chem.* **55**, 2210–2218.

Kok, S. H., Buckle, K. A. and Wootton, M. (1983). Determination of nitrate and nitrite in water using high-performance liquid chromatography. *J. Chrom.* **260**, 189–192.

Lookabaugh, M. and Krull, I. S. (1988). Determination of nitrite and nitrate by reverse-phase high-performance liquid chromatography using on-line post-column photolysis with ultraviolet absorbance and electrochemical detection. *J. Chrom.* **452**, 295–308.

Machackova, I., Zmrhal, Z. and Trckova, M. (1985). Nitrate content of lettuce and spinach leaves in relation to the time of day harvested and the duration and conditions of storage. *Rostlinna Vyroba* **31**, 1151–1156.

Massey, R. C., Key, P. E., Mallett, A. K. and Rowland, I. R. (1988). An investigation of the endogenous formation of apparent total N-nitroso compounds in conventional microflora and germ-free rats. *Food Chem. Toxicol.* **26**, 595–600.

Massey, R. C., Dennis, M. J., Pointer, M. and Key, P. E. (1990). An investigation of the levels of N-nitrosodimethylamine, apparent total N-nitroso compounds and nitrate in beer. *Food Add. Contam.* in the press.

Mertens, J., Van den Winkel, P. and Massart, D. L. (1975). Determination of nitrate in water with an ammonia probe. *Anal. Chem.* **47**, 522–526.

Ministry of Agriculture, Fisheries and Food (1987). *Nitrate, Nitrite and N-Nitroso Compounds in Food.* Food Surveillance Paper No. 20. HMSO, London.

Ministry of Agriculture, Fisheries and Food (1988). *The British Diet: Finding the Facts.* Food Surveillance Paper No. 23. HMSO, London.

Motomizu, S., Kikasa, H. and Toei, K. (1987). Fluorimetric determination of nitrate in natural waters with 3-amino-1,5-naphthalenedisulphonic acid in a flow-injection system. *Anal. Chim. Acta* **193**, 343–347.

Mullins, F. G. P. (1987). Determination of inorganic anions by non-suppressed ion chromatography with indirect ultraviolet absorption detection. *Analyst* **112**, 665–671.

National Academy of Sciences (1981). *The Health Effects of Nitrate, Nitrite and N-Nitroso Compounds.* National Academy Press, Washington.

Norwitz, G. and Keliher, P. N. (1985). Study of the interferences of the spectrophotometric determination of nitrite using composite diazotisation-coupling reagents. *Analyst* **110**, 689–694.

Ohshima, H. and Bartsch, H. (1981). Quantitative estimation of endogenous nitrosation in humans by monitoring N-nitrosoproline excreted in the urine. *Cancer Res.* **41**, 3658–3662.

Olsman, W. J. and van Leeuwen, C. M. (1977). The assay of protein-bound nitrite in meat products. *Zeitschrift für Lebensmittel Untersuchung und Forschung* **164**, 239–242.

Packer, P., Van Acker, B., Reed, P. L., Haines, K., Thompson, M. H., Hill, M. J. and Leach, S. A. (1990) The effect of gastric achlorhydria on the urinary recovery of nitrate in man: relevance to urinary nitrate as a measure of dietary nitrate exposure. *Carcinogenesis* **11**, 1373–1376.

Peattie, M. E., Buss, D. H., Lindsay, D. G. and Smart, G. A. (1983). Reorganisation of the British Total Diet Study for monitoring food constituents from 1981. *Food Chem. Toxicol.* **21**, 503–507.

Pentachuk, J., Haldna, U. and Ilmoja, K. (1986). Determination of nitrate and chloride ions in food by single-column ion chromatography. *J. Chrom.* **364**,

189–192.

Pickston, L., Smith, J. M. and Todd, M. (1980). Nitrate and nitrite levels in fruit and vegetables in New Zealand. *Food Technology in New Zealand (February)*, 11–17.

Raikos, N., Fytianos, K., Samara, C. and Samanidou, V. (1988). Comparative study of different techniques for nitrate determination in environmental water samples. *Fresenius Z. Anal. Chem.* **331**, 495–498.

Rose, D. J., Davies, H. and Parker, N. (1990). *Nitrate and Nitrite — Their Availability and Analysis.* Campden Food and Drink Research Association, Technical Memorandum No. 582.

Ross, N. D. and Hotchkiss, J. H. (1985). Determination of nitrate in dried foods by gas chromatography-thermal energy analyser. *J. Assc. Off. Anal. Chem.* **68**, 41–43.

Rowland, I. R., Key, P. E. and Massey, R. C. (1991). Endogenous N-nitrosation in man assessed by measurement of apparent total N-nitroso compounds in faeces. *Carcinogenesis* submitted.

Saul, R. L., Kabir, S. H., Cohen, Z., Bruce, W. R. and Archer, M. C. (1981). Reevaluation of nitrate levels in the human intestine. *Cancer Res.* **41**, 2280–2285.

Scaife A, Saraiva Ferreira, M. E. and Turner, M. K. (1986). Effect of nitrogen form on the growth and nitrate concentration of lettuce. *Plant and Soil* **94**, 3–16.

Schreiner, G., Kiesel, K. H., Gehlen, K. H. and Fischer, A. (1988). Ionchromatographische methode zur gleichzeitigen bestimmung von nitrit und nitrat in fleischerzeugnissen. *Arch. Lebensmittelhygiene* **39**, 25–56.

Schroeder, D. C. (1987). The analysis of nitrate in environmental samples by reversed-phase HPLC. *J. Chrom. Sci.* **25**, 405–408.

Schuster, B. E. and Lee, K. (1987). Nitrate and nitrite methods of analysis and levels in raw carrots, processed carrots and in selected vegetables and grain products. *J. Food Sci.* **52**, 1632–1641.

Sen, N. P., Young, C. L. and McPherson, M. (1979). Comparison of two extraction procedures for recovering nitrite from cured meat products. *J. Assc. Off. Anal. Chem.* **62**, 1186–1188.

Sherlock, J. and Walters, B. (1983). Dietary intake of heavy metals and its estimation. *Chem. and Ind.* 505–508.

Silva, M., Gallego, M. and Valcarcel, M. (1986). Sequential atomic absorption spectrometric determination of nitrate and nitrite in meats by liquid–liquid extraction in a flow-injection system. *Anal. Chim. Acta* **179**, 341–349.

Standing Committee of Analysts (1982). *Oxidised Nitrogen in Waters 1981.* HMSO, London.

Standing Committee of Analysts (1990). *The Determination of Anions and Cations, Transition Metals, Other Complex Ions and Organic Acids and Bases in Water by Chromatography.* HMSO, London.

Tanaka, A., Nose, N. and Watanabe, A. (1980). Gas chromatographic determination of nitrite in foods as trimethylsilyl derivative of 1 *H*-benzotriazole. *J. Chrom.* **194**, 21–31.

Tanaka, A., Nose, N. and Iwasaki, H. (1982). Determination of nitrate in meat products and cheeses by gas–liquid chromatography with electron-capture detection. *J. Chrom.* **235**, 173–185.

The Preservatives in Food Regulations (1989). *Food Composition and Labelling*, Statutory Instrument (1989) No. 533.

Usher, C. D. and Telling, G. M. (1975). Analysis of nitrate and nitrite in foodstuffs: a critical review. *J. Sci. Food Agric.* **26**, 1793–1805.

Van den Berg, C. M. G. and Li, H. (1988). The determination of nanomolar levels of nitrite in fresh and sea water by cathodic stripping voltammetry. *Anal. Chim. Acta* **212**, 31–41.

Velghe, N. and Claeys, A. (1985). Rapid spectrophotometric determination of nitrate in mineral waters with resorcinol. *Analyst* **110**, 313–314.

Walters, C. L., Smith, P. L. R., Reed, P. I., Haines, K. and House, F. R. (1982). N-Nitroso compounds in gastric juice and their relationship to gastroduodenal disease. In: *N-Nitroso Compounds: Occurrence and Biological Effects*, Eds. H. Bartsch, I. K. O'Neill, M. Castegnaro, M. Okada and W. Davis. International Agency for Research on Cancer, Lyon, pp. 345–355.

Walters, C. L., Gillatt, P. N., Palmer, R. C. and Smith, P. L. R. (1987). A rapid method for the determination of nitrate and nitrite by chemiluminescence. *Food Add. Contam.* **4**, 133–140.

Wootton, M., Kok, S. H. and Buckle, K. A. (1985). Determination of nitrite and nitrate levels in meat and vegetable products by high performance liquid chromatography. *J. Sci. Food Agric.* **36**, 297–304.

Wu, H.-L., Chen, S. H., Funazo, K., Tanaka, M. and Shono, T. (1984). Electron capture gas chromatographic determination of nitrite as the pentafluorobenzyl derivative. *J. Chrom.* **291**, 409–415.

Wu, W. S. and Saschenbrecker, P. W. (1977). Nitration of benzene as method for determining nitrites and nitrates in meat and meat products. *J. Assc. Off. Anal. Chem.* **60**, 1137–1141.

2

Nitrate in water

D. Oakes

2.1 INTRODUCTION

Nitrogen is an essential element for plant growth. In most cases plants satisfy their nitrogen requirement by absorbing ammonium or nitrate from soil water. However, just as water can drain from the soil rather than being taken up by crop root systems, so nitrate in soil solution can leach out of the soil and enter rivers or groundwater supplies. Ammonium is not as mobile and will be held on soil particles, but soil bacteria will rapidly convert it to nitrate. The nitrogen cycle is complex, and is discussed in detail in Chapter 3. However, it is apparent that nitrate leaching from soils has increased overall during the last 20 or 30 years, and has resulted in significant increases in nitrate concentrations in water supplies. Interest in this topic has been heightened by changes in legislation which have resulted in reductions in the nitrate concentration allowed in drinking water supplies. In this chapter the legislation on nitrate in water is discussed, and concentrations in water supplies are reviewed. There then follows a section on the use of models for predicting future nitrate trends, and finally a survey of the options available for controlling nitrate levels in water supplies.

2.2 LEGISLATION ON NITRATE IN WATER

Nitrate in drinking water was first connected with methaemoglobinaemia in infants in 1945. About 2000 cases have been reported worldwide, with a mortality rate of about 8%. Eighty to ninety percent of all cases reported were associated with nitrate concentrations greater than 100 mg NO_3/l. Only 10 cases of the disease have been reported in the UK during the last 30 years. The reliability of much of the reported data on infantile methaemoglobinaemia is questionable, particularly in relation to water quality. Many cases resulted from consumption of water of doubtful microbiological quality. Moreover it is clear that the incidence of the disease is affected by other important factors such as the general health of the population. However, it was considered by the World Health Organization (WHO) and many others that the available evidence was sufficient to justify a nitrate limit on drinking water. In 1970, WHO recommended a level of <50 mg NO_3/l, with a maximum acceptable level of 100 mg NO_3/l. These recommendations were implemented in the UK, and public supplies were kept below 100 mg NO_3/l. When concentrations were above 50 mg NO_3/l the water undertaking informed the medical authorities so that due attention could be given to infant health. The EEC directive on the quality of water for human consumption became law in August 1985. The directive sets a maximum admissible concentration of nitrate in water supplied for human consumption of 50 mg NO_3/l. Under the directive, derogations were allowed where the problem was due to the 'nature and structure of the ground', and there was no danger to public health. In the UK the Department of the Environment decided that derogations for nitrate could be granted, and it proceeded to do so on the advice of the Chief Medical Officer. The maxima allowed in the derogations varied, but the highest concentrations permitted were a three-monthly rolling mean of 80 mg NO_3/l, with no single sample to exceed 100 mg NO_3/l. At the beginning of 1988, the Government announced that existing derogations were being withdrawn. This decision did not reflect any change in medical judgment of health risks; rather, it followed legal advice that the directive did not allow derogations to be given in the particular circumstances which were causing the nitrate limit to be exceeded.

The limits set by various organizations during the past 20 years are summarized in Table 2.1.

2.3 NITRATE CONCENTRATIONS IN WATER

Nitrate concentrations in both surface water and groundwater have been the subject of several reviews during the past decade. The Fourth STACWQ Report (DoE 1984) using data up to 1980 concluded that nitrate concentrations were increasing steadily. It was estimated that there would be further increases of 10–15 mg NO_3/l in mean concentrations in surface waters over the next one or two decades. A slow but persistent increase in nitrate concentrations in some groundwaters was observed, and was expected to continue. The report of the Nitrate Coordination Group (DoE 1986) included additional data up to 1984, and concluded that for surface waters the trends were less certain than those given by the STACWQ report. For groundwaters the additional data confirmed the findings of the STACWQ report. It was also concluded that in regions of low rainfall (parts of eastern and central England) many groundwater nitrate concentrations in unconfined aquifers are likely to exceed 100 mg NO_3/l in the long term. The situation has been brought more up to date in a report by the Department of the Environment (DoE 1989). Overall, there is now less certainty about nitrate trends than previously; some rivers and some groundwater sources continue to display rising nitrate concentrations, whilst in other sources concentrations appear to have stabilized in the last few years.

Table 2.1—Limits for nitrate in drinking water

Organization	Date	Limit (mg NO3/l)	
WHO European Standard	1970	Recommended	<50
		Acceptable	50–100
		Not recommended	>100
WHO International	1971		45
WHO Working Group recommendation	1977	(*a*) *General population*	
		Acceptable	< 50
		Borderline	50–100
		Unacceptable	>100
		(*b*) *Infants* (*below* 6 *months*)	
		Unacceptable	>50
US Environmental Protection Agency	1977	Maximum	45
Health and Welfare, Canada	1978	Maximum acceptable	45
EEC (directive on quality of surface waters intended for abstraction of drinking water)	1977	Imperative limit	50
		Guide level	25
EEC (directive on quality of water for human consumption)	1980	Maximum admissible	50
		Guide level	25
WHO Guidelines on Water Quality	1984	Guide value	45

2.3.1 Groundwater

Groundwater provides about 30% of public water supplies in the UK. However, there is a marked regional variation from over 70% in south-east England to less than 10% in Wales and Scotland. The principal aquifers are the Cretaceous Chalk in southern and eastern England, and the Triassic Sandstone of the Midlands, northern England and south-west England. The Carboniferous and Jurassic Limestones provide about 8% of total groundwater supplies, and are of considerable local importance. The different aquifers exhibit significantly different patterns of nitrate contamination. The main reasons for this are the variability of effective rainfall between different areas and hence between different aquifer types, and the effect of rock properties on the rate of movement of water and nitrate. Effective rainfall (the difference between rainfall and transpiration) may be equated to the amount of water draining from the soil into the underlying unsaturated zone. Lower rates of drainage will provide less dilution for nitrate leached from the soil, giving rise to a preponderance of high nitrate groundwater sources in the drier parts of the UK. Effective rainfall also controls the rate of movement of water and nitrate through the unsaturated zone between the soil and the water table. In the Chalk, rates of downward movement through the unsaturated zone have been estimated to range between 0.5 and 1 m/year (DoE 1986). Movement rates will be greater in Sandstone, and greater still in the well fissured Limestones. The depth of the unsaturated zone will vary over a catchment, giving rise to a commensurate variation in travel times. Travel times through the saturated zone of an aquifer will, in general, be less than through the unsaturated zone, but in the Chalk and Sandstone there is a considerable storage capacity below the water table which will tend to retard the movement of water and nitrate. As a general principle the time of travel of nitrate from the overlying soil to water-supply wells, and hence the aquifer response, may be categorized (Oakes 1990) as in Table 2.2.

Table 2.2 — Travel times from soil zone to water-supply wells

Aquifer	Time (years)	Response
Chalk	10–60	Slow
Sandstone	5–20	Medium
Limestone	0–5	Fast

It should also be realized that some water and nitrate bypasses the slow route in Chalk and Sandstone aquifers by moving through fissures in the unsaturated zone during periods of intense rainfall. It has been estimated that up to 30% of the water and nitrate moving through the unsaturated zone may follow this quick route. In Limestone nearly all of the water and nitrate movement through the unsaturated zone will occur via the fissure network. This means that even in slow-response aquifers, such as the Chalk, there will be a component of rapid response to changes in nitrate leaching from the overlying soils. Some nitrate trends in supply boreholes are shown in Fig 2.1 (Chalk), 2.2 (Sandstones) and 2.3 (Limestone). The apparent differences in nitrate variability between the graphs in each figure are in most cases artifacts of the sampling regimes employed.

The two sources in the Chalk (Fig. 2.1) are in predominantly cereal-growing areas. The trend in Fig. 2.1(a), at Fleam Dyke in Cambridgeshire, was compiled from frequent samples, and is probably typical of the real variability to be found in the Chalk. Concentrations deviate from the mean by up to 10 mg NO_3/l on occasions in an apparently random fashion. The reasons for these deviations are not yet fully understood, but are likely to include aquifer fissuring and layering, variations in effective rainfall and short-term changes in source pumping rate. Fig. 2.1(a) indicates rising nitrate concentrations up to the early 1980s, followed by an apparent equilibration. Prior to 1977, nitrate concentrations rose steadily, though with fluctuations, at an annual rate of about 0.4 mg NO_3/l. The change in trend in the early 1980s has been investigated using mathematical models (Oakes 1990), and almost certainly resulted from a recent change in cereal management and a consequent reduction in nitrate leaching. Although only a small proportion of nitrate leached from the soil moves through fissures in the unsaturated zone, the reduction in nitrate leaching from the soil resulted in a quick response at the water table. The model studies also indicated that a rising trend was likely to be re-established soon. The concentrations in Fig. 2.1(b), at Bircham in Norfolk, are significantly higher, and show a continuing rising trend through the 1980s. The concentrations are higher because the rainfall recharge is less, being about 70% of that at Fleam Dyke. Mathematical modelling of this source showed that nitrate leaching from the soil declined during the 1980s, as at Fleam Dyke. However, in the Bircham catchment there is a much greater range of unsaturated zone depths, resulting in a greater dispersive mixing of nitrate, and a consequent slow but continuing rise in groundwater concentrations through the 1980s. The two sources in the Sandstone (Fig. 2.2) are in areas of mixed farming. At both sites the nitrate concentrations are rising, though with significant variations between years. The rapid changes from year to year are commonly observed in Sandstone sources, and probably result from the heterogeneous character of Sandstone coupled with a low buffering capacity.

Of the two sources in Limestone, that in Lincolnshire is in a cereal-growing area, whilst the source in Gloucestershire is in a mixed-farming region. Nitrate concentrations in the Jurassic Limestone of central Lincolnshire are among the highest in the UK. Low effective rainfall and intensive cereal production on a rapidly responding aquifer have resulted in nitrate concentrations in the outcrop water exceeding 50 mg NO_3/l since the 1960s. The trend shown in Fig. 2.3(a) is typical of the sources in the aquifer. A significant seasonal fluctuation is evident, and demonstrates the rapid response character of the aquifer. The effects of the abnormal rainfall events in 1975/76 and in 1988/89 are also apparent. The abrupt change in trend in early 1980s is again a result of reduced leaching from cereals, but is more pronounced than at Fleam Dyke because all of the nitrate leached moves rapidly down to the water table. Nitrate concentrations at The Swells source (Fig. 2.3(b)) in Gloucestershire are lower than in the Lincolnshire Limestone because leaching losses are lower under the prevailing land use, and the local effective rainfall is significantly higher. Nevertheless, nitrate concentrations are approaching 50 mg NO_3/l and giving cause for concern.

Groundwater nitrate concentrations vary considerably over the UK. The sources shown in Figs 2.1–2.3 are from problem areas where concentrations have already exceeded, or may be expected to exceed in the long term, the maximum admissible value specified in the EC Drinking Water Directive. Concentrations are typically high in the arable agricultural areas, mainly the south and east of England, and low in the mainly grassland areas of the north and west. The rainfall pattern further exacerbates this trend. Fig. 2.4 shows the vulnerability of groundwater to nitrate pollution. The extent of the area within which nitrate concentrations may be expected to exceed 100 mg NO_3/l in the long term was derived from a consideration of farming patterns, hydrogeology, soil characteristics and effective rainfall (DoE 1986). Some groundwater sources outside the defined area have already exceeded or are likely to exceed 50 mg NO_3/l but are not expected to reach levels as high as 100 mg NO_3/l.

2.3.2 Surface water — rivers

Nitrate concentrations in rivers follow a similar geographical distribution to groundwater, and in the south-east of England have been rising for decades. Fig. 2.5 shows

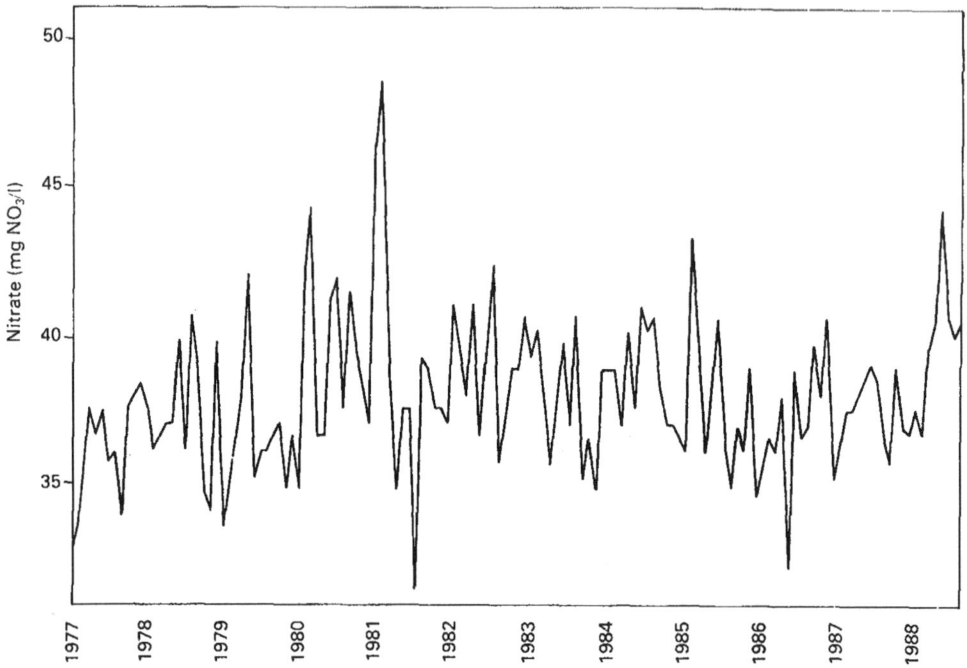

(a)

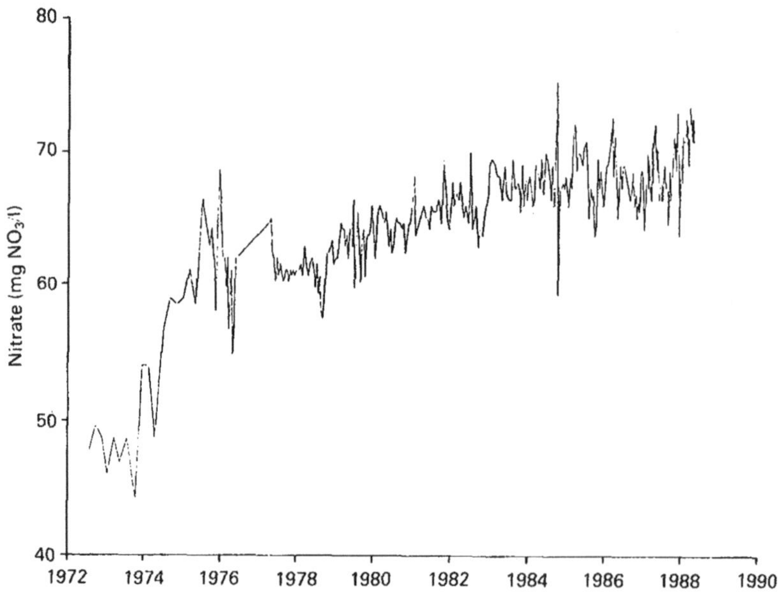

(b)

Fig. 2.1 — (a) Nitrate concentrations in Chalk at Fleam Dyke, Cambridgeshire.
(b) — Nitrate concentrations in Chalk at Bircham, Norfolk.

(a)

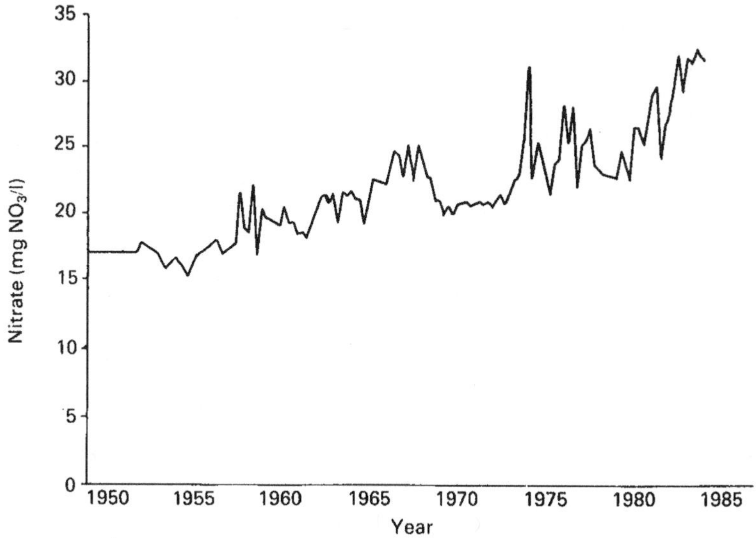

(b)

Fig. 2.2 — (a) Nitrate concentrations in Sandstone at Dotton, Devon.
(b) Nitrate concentrations in Sandstone at Hatton, Staffordshire.

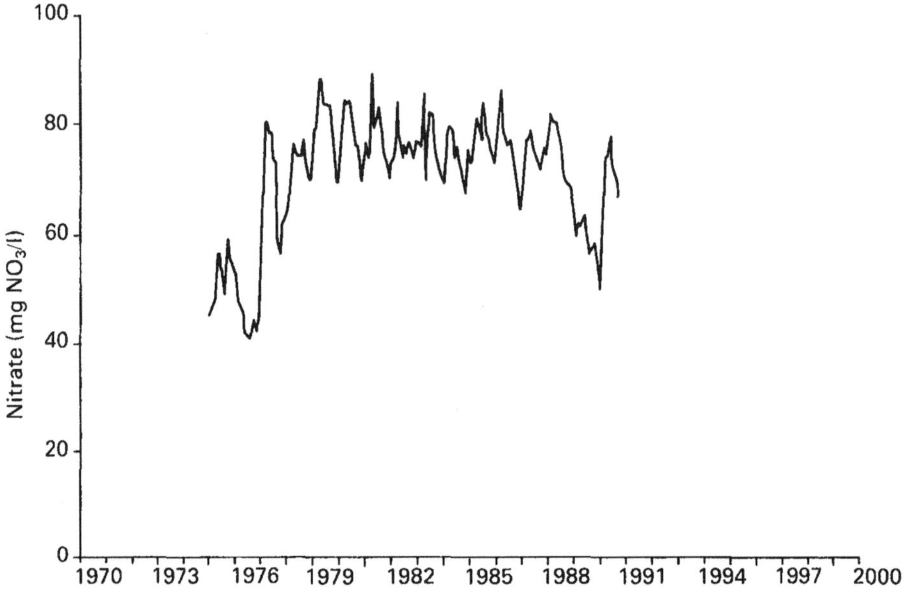

(a)

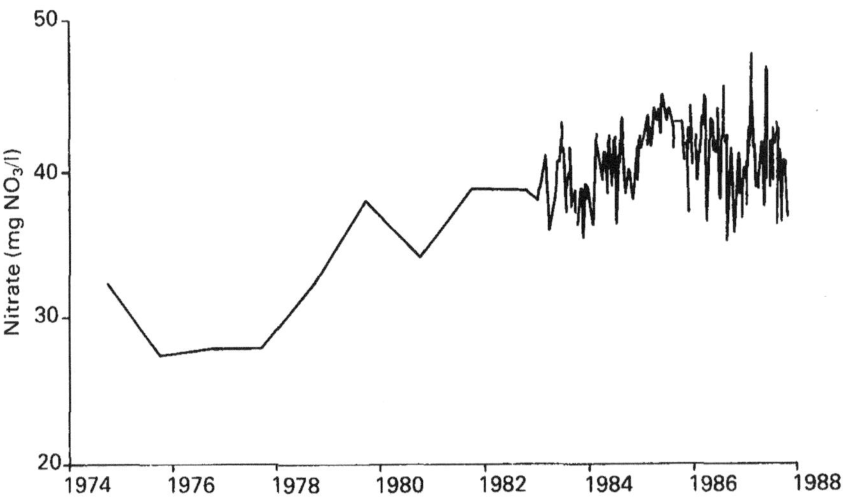

(b)

Fig. 2.3 — (a) Nitrate concentrations in Limestone at Drove Lane, Lincolnshire. (b) Nitrate concentrations in Limestone at The Swells, Gloucestershire.

Fig. 2.4 — Vulnerability of groundwater to nitrate pollution.

five-yearly mean nitrate concentrations at 25 river sampling points in the UK. At nine sampling points, mean nitrate concentrations for 1981–85 were greater than in any previous period. These are primarily in arable farming areas such as East Anglia. In addition to the general rise shown in Fig. 2.5, river nitrate concentrations also show a marked seasonality. Fig. 2.6 shows a time-series of nitrate concentrations in the River Stour at Langham obtained by frequent sampling. It will be noted that while the mean concentration in the period 1976–80 was less than 40 mg NO_3/l (Fig. 2.5), a high of 140 mg NO_3/l was achieved on one occasion, and values in excess of 70 mg NO_3/l were reached in most years. The pattern is typical of arable agricultural areas. Commonly, nitrate concentrations remain low during the summer, and peak at the time of high winter flows. Most rivers in south-east England exceed 50 mg NO_3/l at some time during the winter. In some rivers, nitrate concentrations appear to have stabilized or declined in the past few years. At 15 of the sampling points in Fig. 2.6, mean nitrate concentrations for 1981–85 are below those for the period 1976–80. As with the groundwater trends, the variability of effective rainfall from year to year has a marked effect on nitrate concentrations in rivers. The fall in nitrate concentrations observed in Fig. 2.5 may, in part, have resulted from rainfall fluctuations during the early 1980s. The changes in cereal management implicated in reductions in nitrate concentrations in groundwater during the 1980s have also undoubtedly had an impact on some rivers. In many rivers in the Midlands and the south-east a significant proportion of the total flow is baseflow, so it is not surprising that surface water and groundwater nitrate trends are similar in these areas.

2.3.3 Surface water — reservoirs

Nitrate trends in reservoirs are closely related to the nitrate concentration in the contributing rivers. If the reservoir is large in relation to the inflow then mixing will suppress the large fluctuations in nitrate concentration which are characteristic of many rivers. In addition, nitrate concentrations may be substantially reduced by algal growth and bacterial denitrification. In some reservoirs in the south-east, less than half the nitrate entering via river inputs appears in the reservoir output. The other half is either lost to the atmosphere and sediments, or converted into forms which are removed during treatment.

2.3.4 Conclusions

It will be apparent that nitrate trends in underground and surface waters are difficult to establish, and predictions of further changes have had to be revised at intervals during the past decade. The main reasons for this are:

(a) Mean annual nitrate concentrations are, in general, changing only slowly, and the difference between years is commonly less than the fluctuations within a year.
(b) Land use and fertilizer management are changing and giving rise to changes in nitrate leaching to water supplies.
(c) Meteorological variability can significantly affect nitrate concentrations from year to year and mask any underlying trend.
(d) Changes in source operation will usually affect nitrate concentrations and again obscure the overall trend.

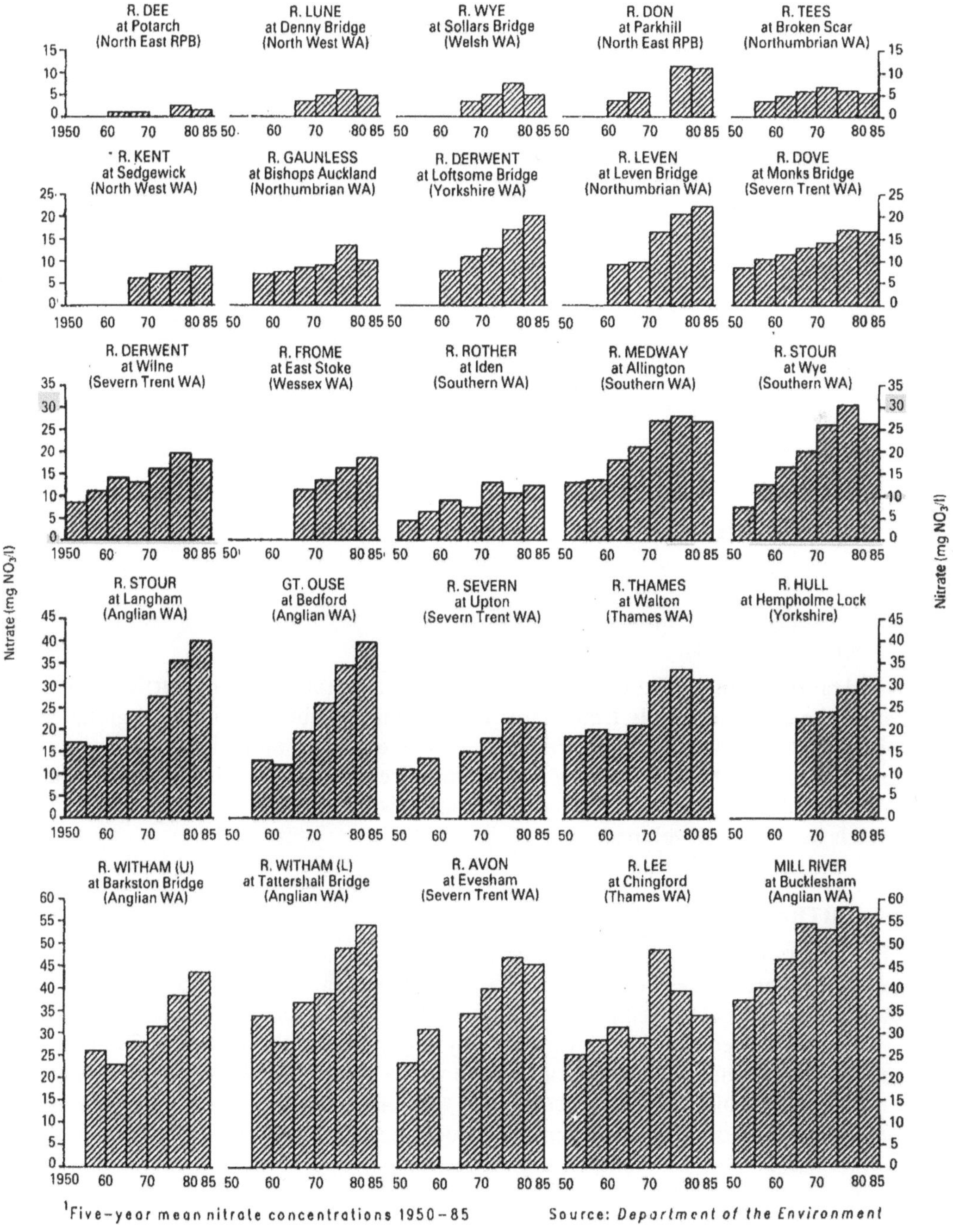

Fig. 2.5 — Five-year mean nitrate concentrations in rivers.

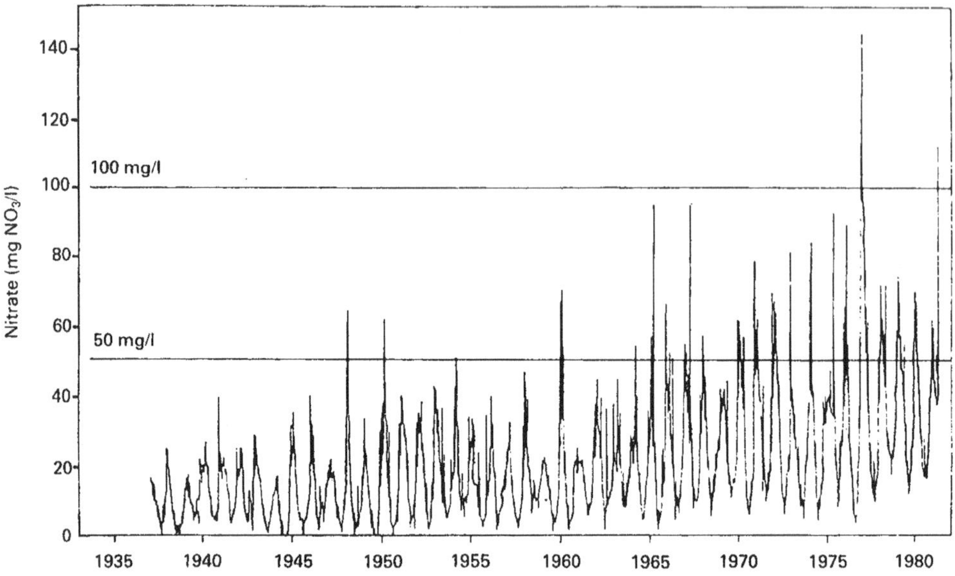

Fig. 2.6 — Nitrate concentrations in the River Stour at Langham

(e) Irregular or infrequent measurement of nitrate concentrations makes trend analysis difficult.

It is clear that firm predictions of future nitrate concentrations in any particular area cannot be deduced simply from statistical information about national, regional or local trends. Such predictions are needed to assess when and where problems of compliance with the EC Drinking Water Directive will occur, so that strategic plans can be made and appropriate actions taken. It is for this reason that a variety of models have been developed.

2.4 NITRATE MODELS

There is a range of models that can be applied to the movement of nitrate from the soil into water. These vary from mechanistic models through to empirical models. Mechanistic models attempt to simulate the physical, chemical and biological processes involved, and are formulated by applying established scientific laws to generate mathematical descriptions of the whole system. Such models have been used to analyse historic nitrate trends and to establish the reasons for the changes which have been observed. In empirical models the detailed internal mechanisms are not simulated, but system output is related to system input in a simple, mathematical form.

2.4.1 Groundwater models

Several models have been developed during the past 15 years to simulate nitrate concentrations in groundwater. Most of these have been mechanistic, and have postulated processes for leaching of nitrate from soils overlying aquifers and for the subsequent transport through the unsaturated and saturated zones (Mercado 1976, Oakes 1982, Mull and Meier, 1989). The main reason for using mechanistic rather than empirical models is that nitrate concentrations commonly show a considerable variation across an aquifer because of variations in land use and in catchment characteristics such as the depth of unsaturated zone. An empirical model would need to be calibrated for each well within the catchment, and could not be calibrated at all for other areas where data were not available. By using a mechanistic model the movement of water and nitrate through the catchment will be simulated, so that concentrations are simultaneously calculated over the entire aquifer. The groundwater nitrate models developed have a lot in common in terms of the processes simulated. The following brief description relates to the WRc model which has been extensively applied to aquifers in the UK. Further details may be found in Oakes (1990). The controlling mechanisms which are represented in the model are the rate of leaching from the soil overlying the aquifer, transport through the unsaturated zone and then through the saturated zone towards pumping wells and rivers. The catchment is represented by a network of cells of square cross-section extending from the ground surface down to the base of the aquifer. In the saturated zone, water and nitrate will move between neighbouring cells under the action of groundwater gradients. A prerequisite to the groundwater nitrate simulation is, therefore, calculation of groundwater levels and flows which may be accomplished with a groundwater flow model. Nitrate leached from the soil zone in each cell is routed downwards through the unsaturated zone to the water table. A dynamic mass balance is then performed in the saturated zone of each cell to calculate the changes with time of nitrate concentration. It is necessary to calculate leaching losses from the soil over the entire catchment area, which may be many 100s of km^2, and over a long historical period because of the considerable delays which may occur during transit through the unsaturated and saturated zones (see Table 2.2). The calculation of nitrate leaching therefore needs to be relatively simple so that both the spatial and the historical aspects may be readily accommodated. In the WRc model this has been achieved by undertaking an extensive programme of drilling and sampling beneath different crops, and relating the nitrate leaching losses measured to crop type and fertilizer-application rate. Table 2.3 shows the leaching-loss relationships which have been inferred. The reduction in leaching from winter cereals for sowings post-1984 was based on the work of Whitmore and Addiscott (1986). They used a computer model of soil nitrate to show that changes which have occurred recently in the management of winter cereals, namely earlier sowing and reduction of autumn nitrogen fertilizer applications, would together reduce leaching by about 40 kg N/ha/h. Movement of water and contained solutes through the unsaturated zone is simulated by a piston-displacement process, driven by water and solute input from the soil into the top of the unsaturated strata. A proportion of the recharge water and nitrate is allowed to bypass the unsaturated-zone matrix, simulating rapid recharge to the water table through fissures. It has been deduced from field measurements that the fissure-flow fraction depends principally on aquifer type, varying from 1.0 to 10% in Chalk and Sandstone and approaching 100% in Limestone. In the saturated zones of the major aquifers' flow occurs predominantly through fissures, though large

Table 2.3 — Nitrate leaching-loss relationships

Land use	Equivalent N leached
Winter cereals	40% N applied[†]
Peas	50 kg N/ha/a
Other arable	50% N applied
Cut grass	10% N applied
Grazed grass	15% N applied
Ploughed grass	280 kg N/ha
Woodland/urban	2–3 mg N/l

[†] reduced by 40 kg N/ha/a for sowing post-1984, with small reductions back to 1978.

volumes of water may be stored in the primary porosity of the rock matrix. In the model it is assumed that water and nitrate move through the fissures at velocities determined by a groundwater flow model. Nitrate and other solutes may diffuse between the fissure water and matrix water under the action of concentration gradients. In general, the concentration in the matrix water will lag behind that in the fissure water, but if nitrate input to the fissure water declines, the diffusion gradient may be reversed. Exchange between the fissure and matrix water exerts a retarding influence on the transport of nitrate through the aquifer.

The computer model has been applied to more than 20 catchments in the UK and has been able to simulate the historic nitrate trends to a satisfactory degree of accuracy. It has been common practice in these studies to obtain historic land-use information from published Parish Statistics, and fertilizer applications from the Survey of Fertiliser Practice (ADAS/FMA).

The results of model application to the Lincolnshire Limestone are shown in Fig. 2.7. The fluctuations in concentration are largely due to the hydrogeological characteristics of the aquifer and the variations in effective rainfall. It is clear from the satisfactory fit of the model to measured concentrations that the mechanisms controlling nitrate in the aquifer are adequately represented. Fig. 2.8 shows the future projection for the same well, with the assumption that agricultural practices, and hence nitrate leaching, will remain the same. In this simulation, future effective recharge was based on historic values so that the important effects of variability could be assessed in terms of likely fluctuations in nitrate concentration. Fig. 2.9 shows the results of model application to a Sandstone catchment in the Midlands. In this case the modelled results shown are annual average concentrations. The model has been used to compare options for controlling further rises in nitrate concentrations. If changes in land use or fertilizer applications are proposed, it is a simple matter to calculate the changes in nitrate leaching which may be expected using Table 2.3. By applying these changes to the model the impact on groundwater nitrate concentrations may be assessed. In Fig. 2.10, predicted average annual nitrate concentrations are shown for two scenarios, one of land use continuing without change and the other of a 220 ha unfertilized grass-protection zone installed around the main supply

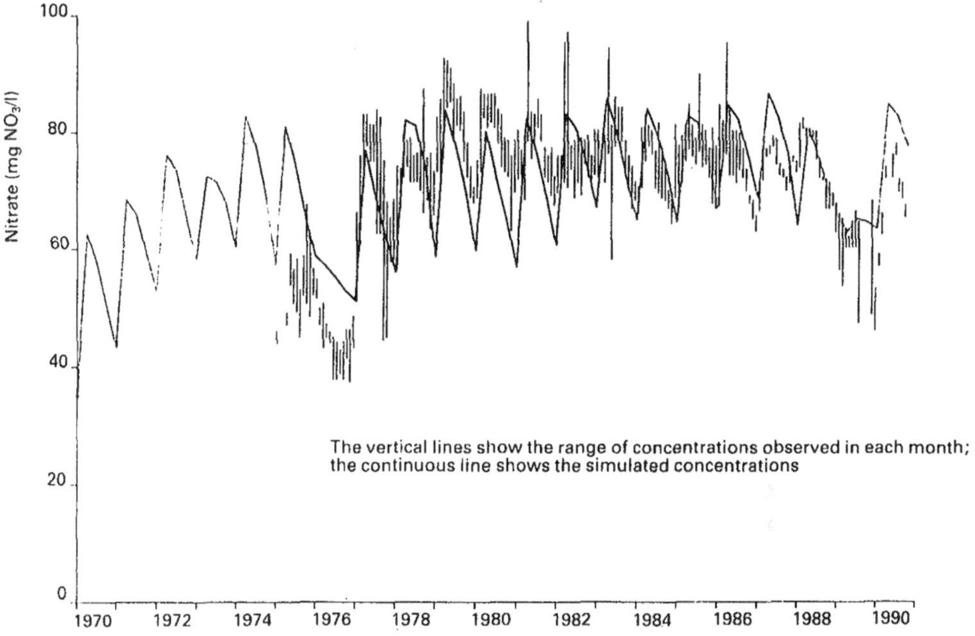

Fig. 2.7 — Observed and simulated nitrate concentrations at Drove Lane. The vertical lines show the range of concentrations observed in each month; the continuous line shows the simulated concentrations.

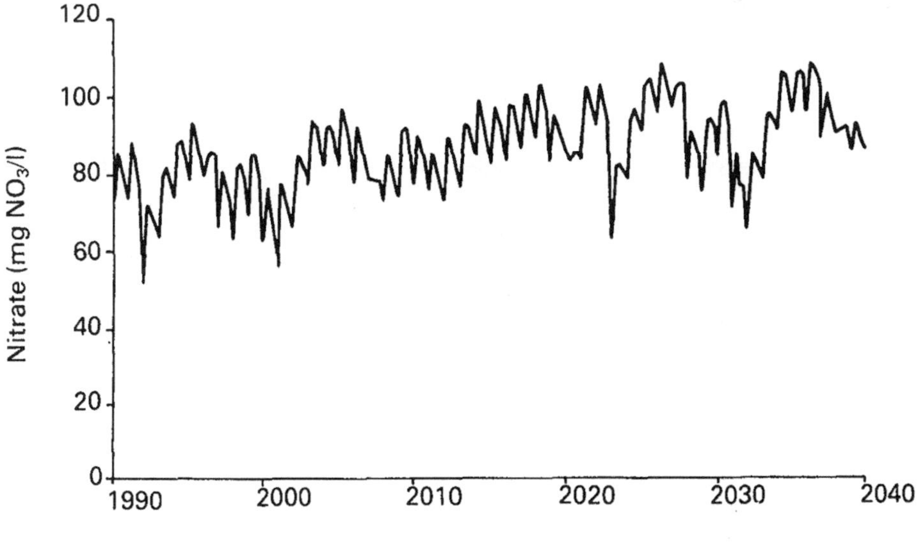

Fig. 2.8 — Simulated future nitrate concentrations at Drove Lane.

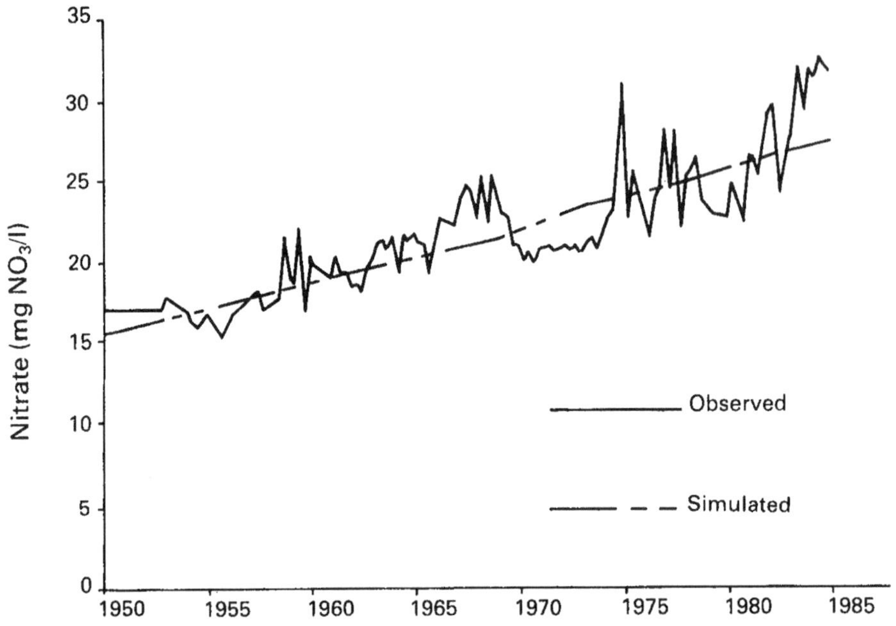

Fig. 2.9 — Observed and simulated nitrate concentrations at Hatton.

borehole in 1990. Other possible control measures which may be readily tested on distributed models such as that described above include catchment-wide land-use changes, fertilizer reductions, borehole relocation and blending of water from different boreholes.

2.4.2 River models

Methods that have been applied to the simulation of nitrate in rivers range from simple time-series techniques through lumped hydrochemical models of catchment behaviour to detailed distributed models that account for spatial and temporal variability.

Time-series methods are suitable where the overall input–output behaviour is of prime importance, and where internal mechanisms are not well understood, but models based on such methods are of only limited use for predicting future trends. Onstad and Blake (1980) developed a model of nitrate concentrations in the River Thames at Walton in which the input, in terms of nitrate available for leaching from the agricultural land of the catchment, was related to the nitrate output from the catchment. The output series was the flow-weighted average annual nitrate concentration in the River Thames at Walton, with an adjustment to remove the relatively small contribution from effluent discharges. The input was the nitrate available for leaching, that is the difference between annual nitrate additions to the catchment from rainfall, soil organic matter mineralization, fertilizers, animal wastes and grassland ploughing, and annual removals in harvested products. Time-series for

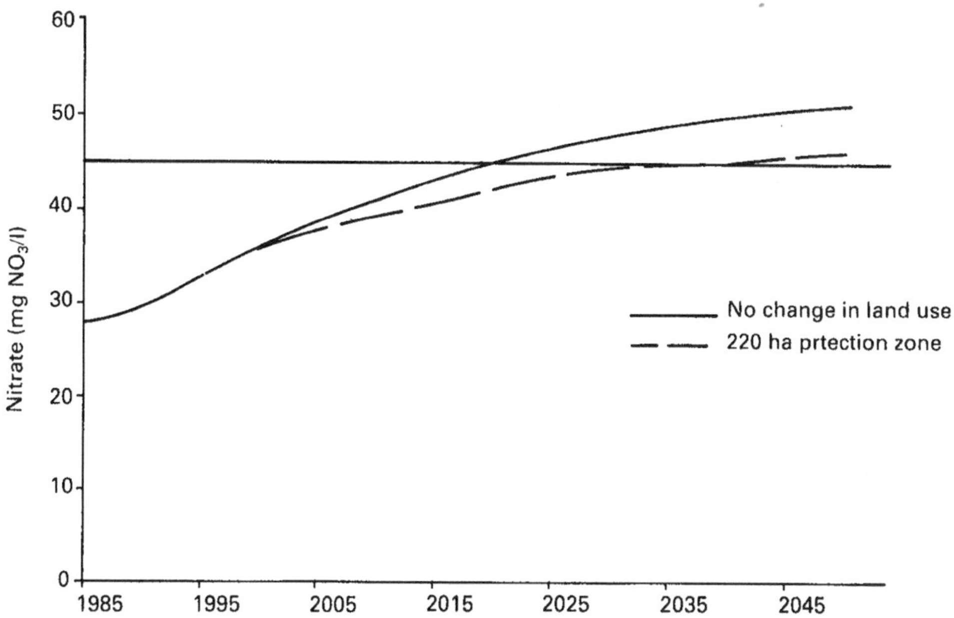

Fig. 2.10 — Simulated future nitrate concentrations at Hatton. Concentrations shown are annual averages; fluctuations about the mean of 5 mg NO_3/l are expected, resulting in an annual average target of 45 mg NO_3/l.

input and output parameters were generated from 1929 to 1975 and related by an autoregressive, moving-average model in the form:

$$c_t=0.124\, c_{t-1}+3.21\, n_t+4.69\, n_{t-7}-1.06$$

where c_t is mean annual concentration in mg NO_3/l in year t, and n_t is annual nitrate available for leaching in year t expressed as a ratio of that available in 1928. The two terms which involve the amount of nitrate available for leaching were considered to represent the effects of surface runoff and groundwater baseflow to the river. The lag of seven years for the groundwater component represents the mean catchment delay time for groundwater flow paths. Comparison between measured and predicted average annual nitrate concentrations is shown in Fig. 2.11. The peak during the late 1940s was identified by the modelling study as arising from extensive grassland ploughing in this period. The sharp rise in concentrations beginning in the 1960s was attributed to increasing use of inorganic fertilizers. The model has been used to predict future trends, but only under conditions of limited change. With empirical models of this type, significant changes in land use, especially if applied in a non-uniform manner, would require recalibration of the parameters.

Such problems of spatial variability can be adequately addressed only by using a distributed model in which the mechanisms of water and nitrate transport are accurately represented. A distributed model was developed for and applied to the

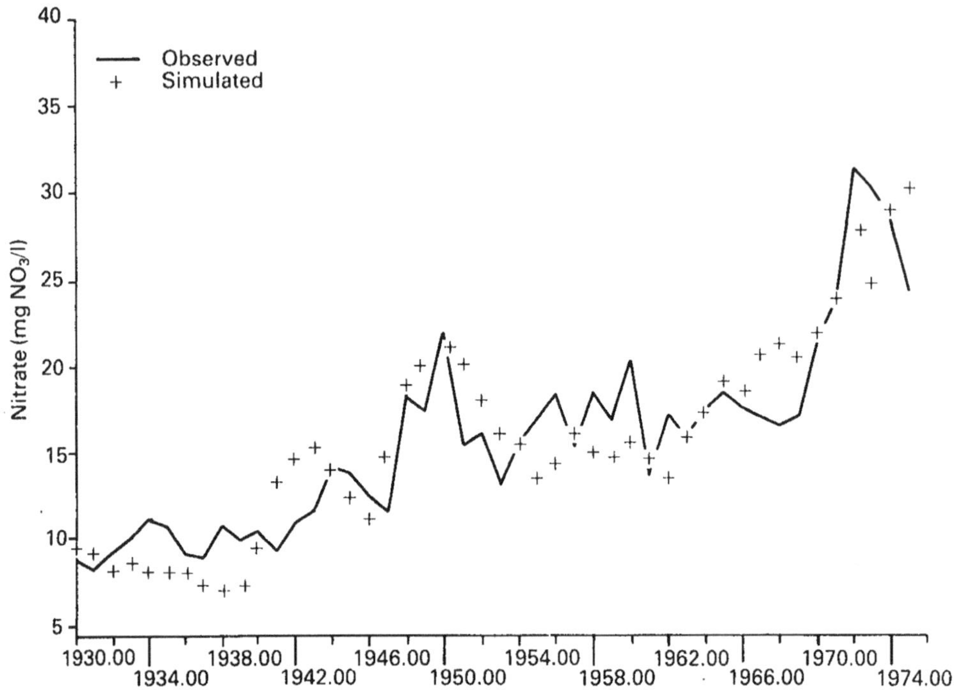

Fig. 2.11 — Observed and simulated nitrate concentrations in the Thames at Walton.

River Thames in a follow-up to the work of Onstad and Blake. In the model (Whitehead and Williams 1984), inputs to the river system from runoff and from groundwater baseflow were calculated within a distributed framework, and used in a river-flow calculation to simulate changes with time in river nitrate concentration. The main problem in application of the model was deriving the nitrate inputs to the river system from runoff and baseflow. In the approach used here, the groundwater component was calculated using the groundwater model described in section 2.4.1. Distributed models were constructed for all groundwater catchments and run with appropriate land-use and fertilizer-application data to provide nitrate concentrations and masses in baseflow discharge to the river system. A similar exercise was undertaken using the leaching calculation of Table 2.3 to estimate the nitrate concentration in runoff. Transport and transformation of nitrate in the river system were modelled by dividing the river network into reaches and undertaking water and nitrate balances within each reach. The nitrate balance included inputs from runoff, baseflow and effluent discharges, and removals by surface-water abstractions and denitrification.Denitrification was represented by a first-order process with a temperature-dependent rate constant. The model was calibrated against observed data. Fig. 2.12 shows a comparison between measured and simulated nitrate concentrations at Datchet over a three-year period. The model has been used to predict future trends, in particular the rate at which river nitrate concentrations are expected to rise, and the related increasing frequency of exceeding 50 mg NO_3/l. The model has also been used to provide nitrate time-series for input to models which simulate nitrate changes in reservoirs.

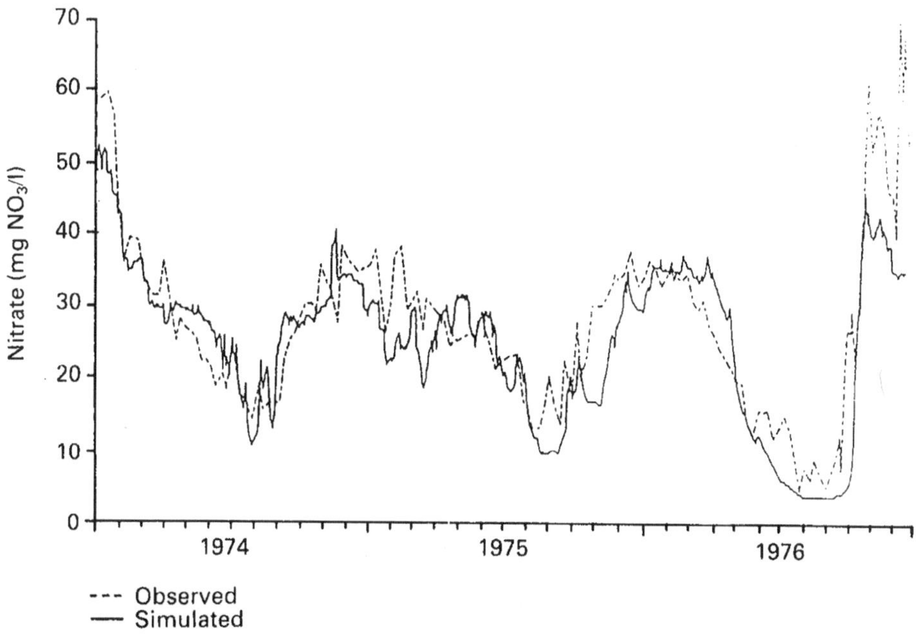

Fig. 2.12 — Observed and simulated nitrate concentrations in the Thames at Datchet.

2.4.3 Reservoir models

Reservoir storage acts as a buffer to fluctuations in nitrate concentration in the input water, and additionally significant denitrification can occur if retention times are sufficiently long. Both processes may be represented mathematically and incorporated in a simulation model. It is common to assume that the reservoir is a fully mixed system so that nitrate concentration in the reservoir is a function only of time. The nitrate mass balance for the reservoir may then be formulated in terms of water and nitrate inputs, the withdrawal rate and the denitrification rate. Rutherford and Davis (1979) used a first-order process with a temperature-dependent rate constant for denitrification to simulate nitrate concentrations in two pumped storage reservoirs. For each reservoir the two-parameter model was calibrated using one year of weekly data, and then validated against other years. Fig. 2.13 shows a comparison between observed and predicted nitrate concentrations in one of the reservoirs for 1976, calibration having been undertaken with 1969 data. Models of this type may be used to predict future trends and to assess the impact on those trends of various operating strategies. Future river nitrate concentrations to input to the reservoir model may be based on historic trends or generated by an appropriate river model.

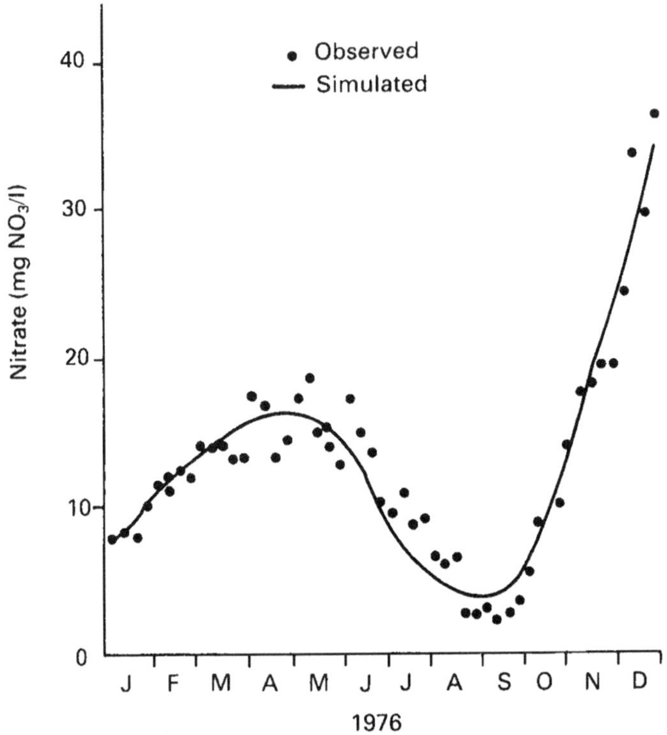

Fig. 2.13 — Observed and simulated nitrate concentrations in Grafham Water.

2.5 NITRATE CONTROL

There is a number of options available to water undertakings for reducing nitrate concentrations in a water supply. The principal options are blending of high-nitrate with low-nitrate water, extended storage, source modification or replacement, and removal by treatment. For small private supplies, point-of-use treatment systems employing ion exchange are available, and may represent the best option for reducing nitrate concentration. However, if not designed, installed and operated correctly these devices can present a potential risk to health. An alternative approach to controlling nitrate concentrations in water supplies is to reduce nitrate leaching from soils by means of land-use changes including restrictions on fertilizer usage. The water industry has always taken the view that nitrate pollution of water supplies has resulted from agricultural activities, and that 'the polluter pays' principle should apply. However, it has been recognized (Harryman 1989) that the situation with regard to nitrate is exceptional for a number of reasons. Firstly, nitrate is a natural and necessary nutrient for plant growth. Secondly, it has been policy over many years to encourage agricultural production and, as part of that, the use of fertilizers. Thirdly, nitrate leaching is a natural soil process, and fertilizer applications play only a part in that process. Finally, there is the additional problem for groundwaters of identifying the polluter, given the lengthy time lag between nitrate leaching from the soil and its appearance in supply wells, which is a common feature of most aquifers.

2.5.1 Blending

Blending supplies is practised widely at present where alternative low-nitrate sources are available within a reasonable distance. Blending of two or more sources will always reduce the nitrate concentration below the highest level in any of the individual sources, and in addition will usually provide attenuation of short-term peak concentrations which are commonly a major cause of concern. Surface waters can be used as low-nitrate sources for blending with high-nitrate groundwaters, but other water-quality problems may occur. As groundwater is generally of very good quality, water undertakings are reluctant to blend it with lower-quality surface water, as higher overall treatment costs may be incurred. As nitrate concentrations continue to rise, the availability of low-nitrate waters for blending with high-nitrate supplies will decrease.

2.5.2 Storage

For surface-water-derived supplies, storage in reservoirs for extended periods can smooth out fluctuations in river water quality, and facilitate nitrate reduction by microbiological activity. There are disadvantages, however, as high-nitrate waters are rich in nutrients and extended storage can lead to severe treatment problems associated with algal blooms.

2.5.3 Source modification or replacement

In most aquifers, nitrate concentrations decline with depth below the water table. Fluctuations in borehole abstraction rates are common, and will result in varying proportions of shallow and deep groundwater being drawn into the borehole so giving rise to significant variations in nitrate concentration. Smoothing of abstraction rates has been observed to reduce nitrate fluctuations, but operational demands may well preclude a change to steady withdrawal conditions. Lining out the upper section of high-nitrate boreholes has resulted in a lowering of concentrations, but at the expense of a significant loss of yield. It is also to be expected that nitrate concentrations will gradually revert to the former high values as abstraction continues and high-nitrate water is drawn deeper into the aquifer. A permanent solution to the nitrate problem may be achieved in some cases by relocating groundwater abstractions to areas where the aquifer is covered by impermeable strata, or to aquifer outcrop areas where leaching losses from the soil are low. Water from aquifers covered by impermeable strata is generally of poor quality, however, and additional treatment may be required. Low-nitrate leaching is typical of woodland and low-productivity grassland. Such land is not common in the nitrate problem areas of south-east England and the Midlands.

2.5.4 Nitrate removal by treatment

Four nitrate-removal processes, namely ion exchange, microbiological denitrification, reverse osmosis and electrodialysis, have been satisfactorily demonstrated at the pilot scale. In the ion-exchange process, water is passed through beds of synthetic resin beads which remove anions including nitrate from the water, exchanging them for equivalent amounts of chloride. When the capacity of the resin becomes saturated the nitrate-removal phase is terminated, and the resin must be regenerated with brine before being returned to service. The regeneration phase produces an effluent, high in chloride, sulphate and nitrate, which must be safely disposed of. Microbiological denitrification employs the same process, which occurs naturally in river and reservoir sediments. Water is passed through a sand or gravel medium which supports a high concentration of bacteria sustained by an energy source such as methanol which is added to the water. Dissolved oxygen is removed by bacteria, and denitrification follows with reduction of nitrate to nitrogen gas. The process produces a waste biomass sludge for disposal. The denitrified water will require reaeration, filtration and disinfection, and so this process is not suitable for surface water sources where conventional treatment facilities will already exist. Membrane processes use selectively permeable membranes to separate ions from treated water. In reverse osmosis, the driving force for separation is hydrostatic pressure to force water through a semi-permeable membrane (largely impermeable to ions) to produce a deionized stream and a concentrated waste stream. In electrodialysis, ions are drawn through an ion-exchange membrane, leaving partly deionized water behind. By using alternative anion and cation permeable membranes, treated water and waste-concentrate streams can be produced.

Construction and operation of denitrification plant will add significantly to the cost of water supply. Booker, *et al.* (1990) estimate unit costs (operating and capital repayment) for 10 mg NO_3/l nitrate removal from a 5 M (million) l/d supply of about 2p per m^3 for both ion-exchange and biological denitrification plant. For a 50 M l/d supply the unit cost falls to about 1p per m^3. With membrane methods the unit cost for both 5 M l/d and 50 M l/d supplies is about 11p per m^3. Membrane methods are currently only competitive for small supplies of up to 1 M l/d. However, membrane technology is developing rapidly, and this situation may change over the next few years.

The South Staffordshire Water Company has recently installed ion-exchange plant at its Little Hay groundwater pumping station to reduce the nitrate concentration of the 3.3 M l/d supply from 67 mg NO_3/l to 25 mg NO_3/l. About 0.2 M l/d of effluent is produced daily and piped 4 km to a Severn Trent Water sewage treatment plant for disposal. It is envisaged that many similar plants will be constructed in the UK within the next decade.

2.5.5 Reduction of nitrate leaching

Just as nitrate leaching has increased over the years as agricultural production has intensified, so nitrate leaching to water supplies can be controlled by means of restrictions on land use and associated fertilizer applications, provided the legal means of implementing the necessary changes exists. Such restrictions could be applied over an entire water catchment, or imposed over a part of the catchment in the vicinity of the water supply source. Areas within which some form of land-use restrictions have been imposed to provide a degree of security to water supplies are commonly referred to as protection zones. The National Rivers Authority (NRA) has a protection policy for such matters as the use of sewage sludge on agricultural land. The policy does not cope well with problems of diffuse pollution such as nitrate from agricultural activities. To control nitrate concentrations in water, protection zones around sources may need to be very large to supply sufficient low-nitrate water to dilute the high-nitrate water from the remainder of the catchment. This is especially so in the lower rainfall areas of eastern England. Croll and Hayes (1988) estimated that in the area of Anglian Water alone, protection zones comprising low-productivity grassland would be needed on about 2000 km^2 of aquifer outcrop to maintain groundwater nitrate concentrations at acceptable levels. In the South and in the Midlands, where rainfall is higher, borehole catchment areas are less extensive, nitrate concentrations are generally lower and, if needed at all, protection zones could be relatively small.

A detailed study of groundwater nitrate control in the Hatton Sandstone catchment in the Midlands was undertaken to develop and test a suitable methodology for catchment assessment, and to investigate the options available (Severn Trent Water 1988). A computer model was developed and used to assess the efficacy of various proposed options for controlling nitrate concentrations in supply wells up to the year 2040 (the planning horizon). All options investigated included blending of different borehole waters because this provides a cost-effective means of reducing the scale of the problem. Other options considered were treatment using ion exchange, crop restrictions, fertilizer reductions and protection zones. In the study, different solutions to the nitrate problem were compared financially by applying cost algorithms to the provision of blending and treatment facilities and to loss of agricultural production. Results from the study and financial comparison of the options are summarized in Table 2.4. Computer predictions for options 1 and 2 are shown in Fig. 2.10.

The costs of solving the problem by water treatment and by installing a protection zone around the main water supply borehole are very comparable given the uncertainties in model predictions and in the cost algorithms. The protection zone option will have additional benefits in terms of providing greater security against other pollutants, such as pesticides.

A wider study of the economic and other consequences of various local options for limiting nitrate concentrations in drinking water was undertaken by DoE (1988). It was apparent that the least-cost solution to nitrate control in any area will depend very much on local factors such as rainfall, land use, geology, current nitrate concentrations, the availability of low-nitrate water for blending and the distance to facilities for disposing of waste effluent from nitrate-removal plant.

Following on from this work a consultation paper (MAFF 1989) was issued in May 1989, setting out the framework for a pilot nitrate scheme. Under the scheme a variety of land-use controls will be tested in selected areas to assess the scope for restricting nitrate leaching into groundwater supplies under different hydrogeological, meteorological and agricultural conditions. Powers are available in the Water Act, which received Royal Assent in July 1989, for these schemes to be introduced on

Table 2.4 — Hatton catchment nitrate-control predictions

	Control option	Impact on nitrate	Present value of costs (£000s)
1.	No change	Nitrate exceeds limit 2015; treatment to reduce nitrate needed from 2015	389
2.	2.2 km^2 protection zone around main borehole, put down to permanent unfertilized grass	Nitrate remains below limit until 2040; water treatment not needed	415
3.	Catchment-wide control — potatoes, sugar beet and oil seed rape replaced by peas; N to cereals reduced by 20%	Nitrate remains below limit until 2040; water treatment not needed	513

a voluntary or mandatory basis. The Government has opted initially for a voluntary scheme, and the National Rivers Authority, in consultation with the Ministry of Agriculture, Fisheries and Food, has identified areas for inclusion in the initial pilot scheme, Fig. 2.14. The scheme involves a three-tier system of voluntary agricultural measures aimed at reducing nitrate leaching. At the first level, an intensive campaign will be implemented in nine areas to advise farmers on ways to avoid unnecessary nitrate leaching. At the second level, 12 regions have been designated as Nitrate Sensitive Areas (NSAs), and in these areas farmers will be compensated for adopting restrictions which go beyond the bounds of good agricultural practice. For inclusion in the scheme, farmers would have to accept upper limits to nitrogen fertilizer, manure and slurry applications, restrictions on grassland ploughing, the use of cover crops in certain circumstances, and the retention of features such as woodland and ponds. At the third level, a range of more severe restrictions, including conversion of arable land to permanent unfertilized grassland, will receive additional compensation. The NSAs have been selected in areas in which nitrate concentrations are approaching the EC Drinking Water Directive limit or exceed it by no more than 20 mg NO_3/l. Areas where the response time to land-use changes is very slow have been excluded. The pilot scheme has initially been planned to last five years. This may be too short, even in the most rapidly responding aquifers, and it is likely that the pilot scheme will need to be extended to obtain adequate data for a comprehensive assessment of the efficacy of control by means of land-use changes. The NRA will monitor the scheme, and assess the effects of the land-use changes implemented on nitrate leaching and on groundwater nitrate concentrations. From the findings

Fig. 2.14 — Pilot nitrate scheme areas.

from the initial pilot scheme, other areas may be designated for inclusion in the scheme. Overall, the scheme will provide valuable new information on the relationships between land use and nitrate leaching, on the veracity of groundwater nitrate models and on the control of nitrate contamination at source.

REFERENCES

Booker, N. A., Hall, T. and Hyde, R. A. (1990). Removal of industrial and agricultural contaminants from groundwater. In: *World Water '89, Proceedings of an ICE Conference.* Thomas Telford.

Croll, B. T. and Hayes, C. R. (1988). Nitrate and water supplies in the United Kingdom. *Environment Pollution* **50**, 163–187.

Department of the Environment (1984). *Standing Technical Advisory Committee on Water Quality,* Fourth Biennial Report, 1981–1983. HMSO, London.

Department of the Environment (1986). *Nitrate in Water.* Report by the Nitrate Coordination Group. HMSO, London.

Department of the Environment (1988). *The Nitrate Issue.* A study of the economic and other consequences of various local options for limiting nitrate concentrations in drinking water. HMSO, London.

Department of the Environment (1989). *Digest of Environmental Protection and Water Statistics.* HMSO, London.

Harryman, M. B. M. (1989). Water source protection and protection zones. *Water and Environmental Management* **3**, 548–550.

Mercado, A. (1976). Nitrate and chloride pollution of aquifers, regional study with the aid of a single cell model. *Water Resources Research* **15**, 731–747.

Ministry of Agriculture, Fisheries and Food (1989). *Nitrate Sensitive Areas Scheme,* MAFF.

Mull, R. and Meier, W. (1989). Nitrate in groundwater. *Proc. TAHS Symposium,* Benidorm.

Oakes, D. B. (1982). Nitrate pollution of groundwater resources — mechanisms and modelling. In: *Non-point Nitrate Pollution of Municipal Water Supply Sources: Issues of Analysis and Control.* Ed. K.-H. Zwirnmann. International Institute for Applied Systems Analysis. Collaborative Proceedings Series CP-82-S4, 207–230. The Institute, Laxenburg, Austria.

Oakes, D. B. (1990). The impact of agricultural practices on groundwater nitrate concentrations. In: *World Water '89, Proceedings of an ICE Conference.* Thomas Telford.

Onstad, C. A. and Blake, J. (1980). Thames basin nitrate and agricultural relations. *Proc. of the ASCE Symposium on Watershed Management,* Boise, Idaho.

Rutherford, J. C. and Davis, J. M. (1979). Modelling nitrate concentrations in pumped storage reservoirs. *Progress in Water Technology* **11**, 327–336.

Severn Trent Water (1988). *The Hatton Catchment Nitrate Study.* Severn Trent Water, Birmingham.

Survey of Fertiliser Practice. Annual Publication. ADAS/FMA.

Whitehead, P. G. and Williams, R. (1984). Modelling nitrate and algal behaviour in the River Thames, *Water Science and Technology* **16**, 621–633.

Whitmore, A. P. and Addiscott, T. M. (1986). Computer simulation of winter leaching losses of nitrate from soils cropped with winter wheat. *Soil Use and Management* **2**, 26–30.

3

Origins of nitrate in water

M. J. Hill

3.1 INTRODUCTION

The nitrate in drinking water comes from a variety of sources, many of which are dependent to some extent on human activity and so vary between regions and with time. The main sources are in rainfall, sewage or animal excreta, biological fixation in the soil and various agricultural practices. However, the relative importance of 'new' sources is being realized continually as more is learned about the nitrogen cycle. Further, the nitrogen from sources other than rainfall or sewage or industrial effluent must be leached from the soil, and the complexity of this process is only slowly being realized as the reasons for the steadily increasing nitrate levels in rivers are slowly elucidated.

Nitrate is the end-product of sewage disposal, because of the microbial breakdown of the nitrogenous compounds in excreta. It is also released from plant material by microbial action, and so an apparently innocuous process, such as ploughing a field of pasture, can have major consequences for the nitrate content of the aquifers served.

Changes in the perceived relative importance of various sources can be deduced from a comparison of Tables 3.1 and 3.2. The two tables compare different things: Table 3.1 is concerned with the origin of nitrogen in groundwater supplies whereas Table 3.2 gives the origins of the nitrogen received by farmland. However, the relative importance of rainfall, nitrogen fixation by microbes and plants and the amount of nitrogen fertilizer would be expected to be similar. In fact, the ratio of fertilizer to biological fixation is consistent whereas the perceived importance of rainfall has increased considerably in more recent studies.

3.2 NITRATE FROM ATMOSPHERIC SOURCES

Oxides of nitrogen are present in the atmosphere from a range of sources, both natural and industrial. Nitrogen may be oxidized to a range of oxides of nitrogen

Table 3.1 — Sources of nitrate in drinking water in North Lindsey (eastern England) (data from Davey (1970))

Nitrogen source	Estimated percentage contribution
Rainfall	2.5
Domestic sewage etc.	10.0
Farm animal excrement etc.	5.0
Applied fertilizer	25.0
Bacterial and plant fixation of nitrogen	7.5
Combined nitrogen in the ploughed layer	50.0

Table 3.2 — The sources of nitrogen received by farmland in the UK (Royal Society study Group 1983)

Source	1978		1981	
	Total	%	Total	%
Biological fixation	0.35	18	0.35	16
Rainfall	0.30	15	0.30	14
Animal feedstuffs	0.20	10	0.20	9
Fertilizers	1.15	57	1.34	61

during lightning discharges. In addition, oxides of nitrogen are formed during the combustion of a number of fossil fuels (including the principal ones of oil, gas and coal); in major conurbations the density of motor traffic ensures that motor exhausts are the major source of atmospheric oxides of nitrogen whilst in industrial areas fuel-intensive factories will be the major sources. These oxides then dissolve in rain water and are further oxidized to nitrous and nitric acids. Photo-oxidation in intense sunlight increases the rate of conversion of nitrogen oxides to nitric acid. However, although nitric acid is a major contributor to acid rain (accounting for 30% of rainfall acidity in northern England) the average nitrate content rarely exceeds 0.2 ppm (Holmes 1974) and the rainfall usually contains more ammonia than oxides of nitrogen. According to Davey (1970), rainfall contributes only 2.5% of the total nitrogen in drinking water. However, measurements of nitrate in association with acid rain show that the concentration is increasing. In accordance with this, a more recent report in 1983 by a Royal Society Study Group (Table 3.2) suggests that it contributes very much more to the total than was thought from the early studies. A contribution of 15% to the total was adduced. This latter conclusion has been reinforced by WHO (1985), who concluded that in some situations the contribution could be as much as 25% of the total load of nitrate, and could well increase in the future.

3.3 BACTERIAL PRODUCTION OF NITRATE IN SOIL

3.3.1 The nitrogen cycle in soil

The nitrogen cycle in soil is complex, but can be portrayed schematically in simplified form as in Fig. 3.1. 'Organic nitrogen in living matter' is principally plant nitrogen,

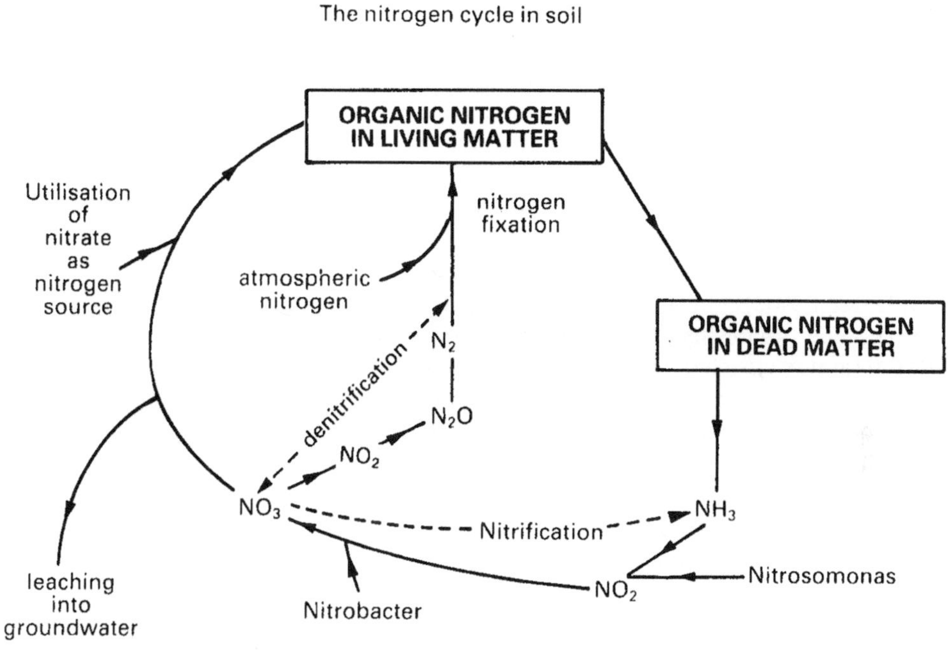

Fig. 3.1 — The nitrogen cycle in soil.

but a small proportion, of course, exists as animal protein (the source of which, however, was originally plant nitrogen). The organic nitrogen in non-living material consists of dead plants and animals together with animal excreta. Animal faeces consist mainly of plant residues (non-nitrogenous) and bacteria; these latter autolyse, releasing bacterial protein which is readily broken down to amino acids (by bacterial proteases and peptidases) which can then be deaminated to release ammonia. The nitrogen present in urine is principally urea in mammals, uric acid in birds and ammonia in reptiles. Bacterial ureases are widely distributed amongst the soil organisms, and this enzyme releases ammonia and carbon dioxide. Bacteria also produce specific uric acid hydrolases which release the ammonia from that substrate. The dead plant and animal tissue is more slowly decomposed but it, too, is eventually mineralized to ammonia.

The oxidation of ammonia to nitrate is a key process in the nitrogen cycle because nitrate is the most readily assimilated form of nitrogen for plants and is by far their major nitrogen source. Under anaerobic conditions (i.e. deeper in the soils), nitrate undergoes denitrification to yield nitrogen gas, but this can re-enter the cycle via the nitrogen-fixing bacteria.

3.3.2 Nitrogen fixation by bacteria

Bacteria able to synthesize nitrate from atmospheric nitrogen were first described by Beijerinck in 1901. They have been named as *Azotobacter spp,* and defined as large rods which are dependent primarily for growth energy on the oxidation of carbohydrates; may be motile with tufted flagella; obligately aerobic and capable of fixing atmospheric nitrogen in the presence of carbohydrate and the absence of combined nitrogen. Beijerinck described two species, namely *Az. chroococcum* and *Az. agilis*, the former of which is widespread in garden soil (or fruitful soil of any sort) and the name of which is derived from the brown pigment possessed by the organism. Since then a number of other species have been isolated from soil and described, including *Az. vitreus*, *Az. vinelandii*, *Az. woodstowni* and *Az. beijerincki.* The organisms fix atmospheric nitrogen and convert it to ammonia, nitrite and nitrate (Beijerinck and van Delden 1902), but only when supplied with a fermentable carbohydrate energy source. The organisms are acid-sensitive in soil, and are unable to flourish below pH 6.0 (and *in vitro* only grow between pH 6.5 and 8.6 (Fred and Davenport 1918).

The nitrogenase has been studied in detail and has been shown to require a cofactor containing molybdenum and iron. This rationalized the well established fact that nitrate builds up in soils deficient in molybdenum. The enzyme consists of two proteins (both oxygen sensitive), one containing molybdenum and iron (Mo–Fe-protein) and the other only containing iron (Fe-protein). When the two proteins are combined in the presence of MgATP and a suitable electron donor (such as ferrodoxin) they catalyse the reduction of N_2 to NH_3 and of H^+ to H_2.

There is now known to be an alternative pathway for nitrogen fixation in which the Mo–Fe-protein is replaced by a vanadium-dependent protein (V–Fe-protein). The significance of this alternative pathway has still to be established. The properties of the molybdenum-based nitrogenase have been reviewed by Orme-Johnson (1985) and by Eady (1986); the evidence for the alternative pathway was given by Bishop *et al.* (1980), and the two pathways have been reviewed by Eady *et al.* (1988) and by Yates (1988).

Early studies of the nitrogen-fixing bacteria included in the family Azotobacteraceae the genera *Azotobacter* and in addition the genera *Azomonas*, *Beijerinckia*, and *Derxia.* However, as knowledge of the organisms has increased, there has been considerable unease about grouping these organisms together taxonomically. Thompson and Sherman (1979) concluded that the family of Azotobacteria is composed of 'a collection of bacterial genera that fix nitrogen nonsymbiotically, that are Gram-negative, nonspore-forming etc and that cannot be readily identified with genera in other families'. This family is now thought to include only *Azotobacter* and *Azomonas*, with *Beijerinckia* and *Derxia* transferred elsewhere. *Derxia* is able to fix nitrogen autotrophically using H_2 as the acceptor (Pedrosa *et al.* 1980); *Azotobacter* cannot do this. The characteristics of the free-living nitrogen-fixing bacteria are summarized in Table 3.3.

Table 3.3 — The nitrogen-fixing bacteria found free living in soil and water and in the rhizosphere of legumes

	A	B	C	D	E	F	G
Nitrogen fixation — aerobic	+	+	+	+	−	−	−
— anaerobic	−	−	−	−	+	+	+
Root-associated fixation:							
without root hypertrophy	Some	−	−	−	+	−	−
with root hypertrophy	−	−	−	−	−	+	−
Cell morphology:							
ovoid to rod shaped	+	+	−	−	−	−	−
dumbbell shaped	−	−	+	−	−	−	−
rod with granules	−	−	−	+	−	−	−
simple rods	−	−	−	−	−	+	+
spiral	−	−	−	−	+	−	−
Motility	Some	+	+	+	+	+	−

Key: A=Azotobacter; B=Azomonas; C=Beijerinckia; D=Derxia; E=Azospirillum; F=Rhizobium; G=Klebsiella.

In addition to the free-living *Azotobacter*, nitrogen fixation can also be carried out by *Rhizobium* spp when growing in association with root nodules on leguminous plants. These organisms were first described by Beijerinck (1888) and are defined as minute rods, motile when young; obligately aerobic; able to fix nitrogen in the presence of fermentable carbohydrate and the absence of compounds of nitrogen; producing nodules on the roots of leguminous plants. The organism can also be found free-living in soils and waters of a range of origins. Bewley and Hutchinson (1920) described a cycle of changes in the organism, starting with (1) a pre-swarming, non-motile stage, when the organisms are minute, followed by (2) a second pre-swarming stage, when the organisms are still non-motile but are twice as large and coccoid, followed by (3) a swarming stage, when the organisms are highly motile and extremely small (less than 0.2μ wide) ellipsoids, then (4) a rod stage, with the organisms still motile, and finally (5) a highly vacuolated stage. Calcium phosphate causes the change from pre-swarming cocci to rod shapes; acid conditions favour to formation of vacuolated forms whilst alkaline conditions in soil favour vigorous growth and multiplication with no change in morphology. Within the nodules the organisms are rod-shaped and may form filaments or pleomorphic forms.

Fred and Davenport (1918) studied the properties of *Rhizobium* spp isolated from nodules of different legumes, and found differences, suggesting that the organism could adapt to life in symbiosis with the different plants. Strains have a distinct host specificity and then enter a state of symbiosis with the host plant. Unlike the *Nitrobacter*, the *Rhizobium* spp are heterotrophs and freely ferment carbohydrate principally by the Entner–Doudoroff pathway or the pentose cycle (Keele *et al.* 1969, Ronson and Primrose 1979).

3.3.3 Nitrate from soil organic nitrogen

It was first demonstrated 100 years ago by Winogradsky (1890) that nitrate is synthesized from nitrogenous compounds in soil and water. There is relatively little information on the metabolism of complex organic nitrogen to ammonia, but there is a great deal of information on the subsequent metabolism of the inorganic nitrogen. Under conditions that favour oxidation, ammonia is oxidized to nitrite then nitrate, whilst under anaerobic conditions denitrification occurs as the nitrate is reduced to nitrite then either ammonia or nitrogen gas. Photosynthetic plants derive most of their nitrogen from nitrate, and so nitrate released from organic nitrogen material in the soil is available for uptake by plants, to be converted back to organic nitrogenous compounds, or may be leached from the soil by the downward movement of water and its dissolved solutes.

In a series of papers in the 1890s, Winogradsky demonstrated that nitrification includes two major steps, namely (a) the oxidation of ammonia to nitrite (carried out by *Nitrosomonas* spp) and (b) oxidation of nitrite to nitrate (carried out by *Nitrobacter* spp). This remained as essentially the level of our understanding of nitrification until the 1950s, since when there has been an intensive study of both nitrification and of denitrification.

The process of nitrification can be divided into (a) the mineralization of organic nitrogen to yield ammonium ions, (b) the oxidation of ammonium ions to nitrite, and (c) the oxidation of nitrite to nitrate. Denitrification requires different metabolic pathways and different conditions. The major organisms identified as capable of steps (b) and (c) of nitrification are listed in Table 3.4. All are chemoautotrophic organisms (i.e. derive both their carbon and nitrogen requirements from inorganic sources such as CO_2, CH_4, NH_3, NO_2, N_2).

3.3.3.1 Deamination of soil nitrogen compounds

As has already been noted, a very high proportion of the nitrogen in soil is fixed in complex nitrogenous compounds that are unavailable to plants or to leaching into the groundwater. Such complexes are broken down at the same time as the general microbial breakdown of dead and decaying plant material in soil, and by processes that are optimized during the composting of garden waste. Such breakdown is accomplished by consortia of microorganisms including bacteria, protozoa, yeasts, fungi, ciliates, etc. This is the main source of available nitrate in the topsoil (the top 25 cm, according to Addiscott (1988)), and it has been estimated by Addiscott and Powlson (1989) that there is 5000 kg of organic nitrogen per hectare in the soil (and between 2000 and 15 000 kg per hectare according to the Department of the Environment (1988). Approximately 1–3% of this is mineralised to ammonium salts annually by the consortia of soil organisms.

Table 3.4 — Differentiation of the *Nitrobacter*

	A	B	C	D	E	F	G	H	I
NH_3–NO_2	–	–	–	–	+	+	+	+	+
NO_2–NO_3	+	+	+	+	–	–	–	–	–
Shape — straight rod	+	+	–	–	+	–	–	–	–
coccus	–	–	+	–	–	+	–	–	–
helical	–	–	–	+	–	–	+	–	–
other	–	–	–	–	–	–	–	+	+
Reproduction									
binary fission only	–	+	+	+	+	+	+	+	+
budding	+	–	–	–	–	–	–	–	–
Habitat — soil	+	–	–	+	+	+	+	+	+
sewage	+	–	–	–	+	+	+	–	–
water	+	–	–	–	+	+	+	–	–
seawater	+	+	+	+	+	+	–	–	–

Data from Bergey (1989). *Key:* A=Nitrobacter; B=Nitrospina; C=Nitrococcus; D=Nitrospira; E=Nitrosomonas; F=Nitrosococcus; G=Nitrosospira; H=Nitrosolabus; I=Nitrosovibrio.

3.3.3.2 Oxidation of ammonium to nitrite ions

The organisms in this group (termed Nitrosomonas spp by Winogradsky (1890)) obtain their energy by the oxidation of ammonia to nitrite and use CO_2 as their carbon source. The rate of assimilation of CO_2 is quantitatively related to the amount of ammonia oxidized (Winogradsky 1890).

Nitrosomonas spp are defined as rod-shaped organisms, often motile with polar flagella, which derive their energy from the oxidation of ammonia to nitrite. *In vitro*, this ammonia oxidation is inhibited under certain conditions by the presence of organic matter and occurs maximally in purely inorganic media containing ammonium salts and a source of carbonate. *Nitrosomonas* is one of the few organisms able to synthesize organic matter from purely inorganic materials without the use of photosynthetic pathways. Its optimum growth temperature is 25–30°C, and it requires well aerated conditions; the optimal pH for growth is 7.5 to 8.0 and the optimum substrate concentration is 2–10 mM. The organism is inhibited by light.

Organisms other than the *Nitrosomonas* spp of Winogradsky are able to oxidize ammonia to nitrite (Table 3.4). Cutler and Mukerji (1931) isolated a number of organisms from soil, most of which were Gram-positive, non-sporing, non-motile, strictly aerobic rods able to grow in the presence of organic matter. Whereas *Nitrosomonas* spp requires alkaline conditions, the organisms described above were able to form nitrite at a range of pH values between 4.8 and 7.3. In addition, Winogradsky and Winogradsky (1933) described two further ammonia-oxidizing genera, namely *Nitrosospira* and *Nitrosocystis* (later renamed as *Nitrosococcus*) isolated from soil.

Krummel and Harms (1982) studied the effect of the presence of various organic substrates on ammonia oxidase activity; they found that whereas organic matter inhibits the activity of *Nitrosococcus* it had no such effect on *Nitrosomonas*, and they actually observed stimulation of the growth of a strain of *Nitrosospira* by acetate and pyruvate.

Suzuki (1974), in his review of inorganic oxidation, proposed that the oxidation of ammonia followed the scheme:

$$NH_3+O_2+AH_2 \rightarrow NH_2OH+H_2O+A \quad (1)$$

$$NH_2OH+2cyt.\ c\ (Fe^{3+}) \rightarrow NOH+2H^++2\ cyt.\ C\ (Fe^{2+}) \quad (2)$$

$$2\ cyt.\ C\ (Fe^{2+})+\tfrac{1}{2}O_2+2H^+ \rightarrow 2\ cyt.\ c\ (Fe^{3+})+H_2O \quad (3)$$

$$NOH+A+H_2O \rightarrow NO_2+AH_2+H^+ \quad (4)$$

Hollocher *et al.* (1981) showed that the oxygen for reaction (1) comes from molecular oxygen, whilst the oxygen for reaction (4) is from water (Anderson and Hooper 1983). Reactions (2)–(4) are the energy-yielding part of the sequence and involve cytochrome C and P-460 (Hooper *et al.* 1978). Whereas reaction (1) tends to be affected by the presence of organic compounds, the energy-yielding part of the sequence is unaffected by such compounds (Hooper and Terry 1973).

Because all of the intermediates are extremely unstable they are thought to be enzyme-bound. The oxidation of hydroxylamine to nitrite (reactions (2)–(4) is carried out by intact cells of *Nitrosomonas* or *Nitrosococcus*, whilst cell extracts require the presence of an electron acceptor identified as cytochrome c. The enzymes identified are termed ammonia oxidase (responsible for reaction (1)) and hydroxylamine oxidoreductase (responsible for reactions (2)–(4). This latter has been further separated into hydroxylamine dehydrogenase (reaction (2)) and nitrite synthetase (reaction (4)). Cells of *Nitrosomonas* are extremely rich in cytochrome and this gives them a red pigmentation (Dispirito *et al.* 1985).

Very little is known of the ammonia oxidase. It is likely that the substrate is ammonia rather than the ammonium ion; the enzyme may be a metallo enzyme and may require the involvement of a haemoprotein and possibly cytochrome P-460. This cytochrome also appears to be involved in the hydroxylamine dehydrogenase reaction.

Nitrosomonas is ubiquitous and is found in a wide range of soils, not only those associated with sewage or animal manure or soils treated with them. Sims and Collins (1959) found *Nitrosomonas* in Australian desert soils, for example.

Nitrosolabus is found in arable or grassland soils but not forest soils and is very resistant to desiccation and storage. Evidence is accumulating that *Nitrosolabus*, not *Nitrosomonas*, is the dominant nitrifier in agricultural soil (Walker 1978). *Nitrosospira*, in contrast, is dominant in the (acidic) Bangladesh tea soils. Further, it has recently become apparent that some heterotrophic organisms such as *Arthrobacter* spp and some fungi (e.g. *Aspergillus* spp) may be the primary source of nitrification in some soils (Killham 1986). The recent evidence has been reviewed by Kuenen and Robertson (1988).

3.3.3.3 Oxidation of nitrite to nitrate

Soil organisms carrying out this step were collectively termed *Nitrobacter* spp by Winogradsky (1891), and more recent steps to identify the range of species within this group have so far failed to subdivide this group to any great extent. A wide range of heterotrophic as well as chemoautrophic organisms has been identified as being able to carry out this step under controlled laboratory conditions but, for example, in the study by Fliermans *et al.* (1974) fifteen apparently distinct isolates of organisms able to oxidize nitrite to nitrate under normal conditions and isolated from a wide range of environments (e.g. soil, sewage, water, etc.) all proved to be either *Nitrobacter winogradskyi* or the closely related *Nitrobacter agilis*. Although *Nitrococcus* and *Nitrospira* have been reported to oxidize nitrite to nitrate, both have only been reported in marine sediments and have not been found in soil or sewage. This lack of diversity of organisms carrying out a reaction that is of key importance in the nitrogen cycle is puzzling.

Nitrobacter spp are defined as rod-shaped non-motile organisms not growing readily on organic matter or in the presence of ammonia, and able to obtain their energy from the oxidation of nitrite to nitrate. The inhibition of growth by ammonium ions, so readily apparent *in vitro*, is much less apparent in soil. The organism uses CO_2 or carbonate as the carbon source.

The oxidation of nitrite to nitrate involves the participation of cytochromes a_1 and a_3. The generation time of *Nitrobacter* is slow, being reported to be in the region of 13 hours even under optimal growth conditions *in vitro*. The organisms are killed by direct sunlight. The nitrite oxidase is an induced enzyme, its production being suppressed by growth *in vitro* in the presence of amino acids, pyruvate or acetate.

The detailed mechanism of nitrite oxidation has been studied in detail and is thought to be a three-stage process:

$$NO_2^- + H_2O + 2\ \text{cyt.a}\ (Fe^{3+}) \rightarrow NO_3^- + 2H^+ + 2\ \text{cyt.a}\ (Fe^{2}) \qquad (1)$$

$$2\ \text{cyt.a}\ (Fe^{2+}) + 2\ \text{cyt.c oxidase}\ (Fe^{3+}) \rightarrow 2\ \text{cyt.a}\ (Fe^{3+}) + 2\text{cyt.c oxidase}\ (Fe^{2+}) \qquad (2)$$

$$2\ \text{cyt.c oxidase}\ (Fe^{2+}) + 2H^+ + \tfrac{1}{2}O_2 \rightarrow H_2O + 2\ \text{cyt.c oxidase}\ (Fe^{3+}) \qquad (3)$$

The oxygen source for the overall reaction was shown to be H_2O by Kumar *et al.* (1983) and Hollocher (1984). Tanaka *et al.* (1983) showed that the electron acceptor for NO_2^- is cytochrome a_1.

3.3.3.4 Denitrifying pathways

Whereas nitrification is the dominant pathway of nitrogen metabolism in the upper (and aerobic) levels of soil and water, in the deeper anoxic regions the reverse denitrification process is important. The main sequence is (Payne 1973, Pichinoty 1973):

$$NO_3 \rightarrow NO_2 \rightarrow NO \rightarrow N_2O \rightarrow N_2$$

The first step is carried out by nitrate reductase, an enzyme that is widely distributed amongst microbes. The activity and action of the enzyme is dependent on the oxygen, tension; in the presence of oxygen, denitrifying bacteria derive their energy from oxidative pathways whilst in the absence of oxygen they use nitrate reductase as an energy source and as a terminal electron acceptor. Nitrate reductase is produced by a range of enteric organisms classified as nitrate assimilators (i.e. they reduce nitrate to nitrite then to ammonia which is then incorporated into bacterial protein); soil organisms tend to be nitrate dissimulators or denitrifiers which reduce nitrate through to nitrogen but only as an energy source and an electron sink, and do not integrate the nitrogen into biomass. Nitrate reductase is a molybdo-iron-labile sulphide protein (Forget 1971).

The second step is nitrite reduction which, with denitrifying bacteria, yields nitric oxide (Payne 1973, Pichinoty 1973). When soil populations are incubated at acid pH in the presence of nitrite, nitric oxide is given off (Garcia 1973) and whole cells or extracts of known denitrifiers do the same. For example, a fraction from a *Pseudomonas* strain extract stoichiometrically reduced nitrite to nitric oxide (Payne *et al.* 1971).

Nitric oxide has been demonstrated to be reduced further to nitrous oxide; the reaction is suppressed by excess nitrate or nitrite (Payne and Riley 1969), suggesting a control mechanism which permits the preferential reduction of nitrate or nitrite (Payne and Balderston 1978).

Nitrous oxide reductase activity can be induced by the anaerobic growth of denitrifiers in the presence of nitrate, nitrite or nitrous oxide, and reduces nitrous oxide to nitrogen gas (Payne 1973). The enzyme is inhibited by acetylene.

In addition to the denitrification pathways yielding nitrogen gas, there are also assimilatory and dissimilatory pathways for the reduction of nitrate through nitrite to ammonia. This has been reviewed recently by Cole (1988). Again, the enzyme has a molybdenum cofactor and so deficiency of molybdenum in the soil results in a build-up of nitrate.

3.3.4 Nitrate from plant sources

Conversion of the nitrogenous components of plants is a major factor in freshly ploughed fields. Old grassland, when ploughed, can yield up to 4000 kg nitrate per hectare during the next 18 years (Addiscott 1988).

It is estimated that since 1950 in the Federal Republic of Germany about 1 million hectares of grassland has been ploughed and transformed into arable land. This has been estimated to have led, through mineralization, to the formation of three times as much nitrate as has been applied as nitrate fertilizer during 1980. Following the ploughing of grassland in south-east England (a chalk soil region), the nitrate enrichment of the soil was due mainly to the mineralization of the organic matter (Foster *et al.* 1982).

Nitrogen fixation by legumes is important because the nitrate generated is released in part into the soil, particularly after the plants have died Furrer 1986).

3.3.5 Nitrate from animal excreta and sewage

Nitrate is the end-product of sewage disposal and results from the microbial breakdown of the organic nitrogen. In older type sewage works the sewage was, effectively, spread on soil and allowed to percolate through, undergoing oxidation as it proceeded. The capacity of such a sewage works was, of course, limited. In the modern sewage works, the material is stirred as slurries in ponds and aerated to give a maximum rate of digestion. The conditions in such slurries are unfavourable to the faecal microbes and so the slurries must be maintained both until the faecal components have been digested and until the faecal bacteria are no longer detectable. Obviously a proportion of the nitrogen from the sewage goes into the bacterial fraction as organic nitrogen and is disposed of in the sewage sludge; however, the major part remains in the effluent to be discharged into suitable rivers. In the UK more than 95% of the population lives in dwellings connected to public sewers, and so the contribution of sewage to the nitrate load in groundwaters is stable nationally and is unlikely to increase greatly over the coming decades. The problem of nitrate from sewage can, however, be great in rivers or groundwaters close to large urban conurbations and where there are limited alternative routes for sewage effluent discharge.

In many countries, human sewage is used, like animal excreta, as manure to be spread on the fields. As with farmyard manure, human sewage provides a readily available nitrogen source to the soil organisms and is rapidly converted to nitrate. If there is good crop cover on the soil at the time of spreading the sewage, the nitrate will be taken up efficiently by those crops (see, for example, Table 3.5) but if the sewage is spread in the autumn it can aggravate the problem of nitrate leaching into the groundwater.

Table 3.5 — Effect of cropping on the nitrate nitrogen runoff to groundwater (Christy *et al.* 1973)

Cropping system	Nitrogen fertilizer added (kg/ha)	Nitrate nitrogen runoff (kg/ha)†
Fallow, ploughed	0	0.91
Maize, oats	0	0.34
Continuous maize	10	0.10
Continuous maize	200	0.01

† The runoff resulting from rainfall of 11.4 cm in one month (June).

Organic manure can be applied in solid or liquid (slurry) form. The solid form, because of its bulk and solidity, can only be applied to ploughed fields since otherwise there is a risk of crop damage. It is very rapidly mineralized to nitrate, and severe leaching occurs during the winter if it is applied during the autumn. In consequence it is not efficiently utilized by the crops (Dutchion 1985) so farmers are recommended to delay application until early spring when its utilization can be 70% effective (Van Dijk and Sturm 1983) compared with 30% when applied in winter (Vetter and Steffens 1983). This is discussed in more detail in the next section.

During recent decades, with the spread of 'factory farming' there has been an increase in the numbers of animals and therefore in the amount of manure produced, and this has led to localized problems on manure disposal. In such areas the dumping of manure has caused serious nitrate leaching. With cattle, almost 80% of the nitrogen consumed in feed is excreted. Thus when cattle are raised on grassland treated with fertilizer to improve grass growth (thereby allowing more animals per hectare), and since the cattle are often given supplementary concentrates, the increased amount of manure deposited on the land causes excess net input of nitrogen which on some soils can give a rate of nitrate leaching similar to that seen on arable land (Cooke 1984, Ryden *et al.* 1984).

3.4 NITRATE FROM FERTILIZER

Nitrate may be applied in large amounts to arable land. In 1957 the Ministry of Agriculture recommended that between 75 and 90 kg of nitrogen should be applied per hectare for winter-sown wheat production. It was estimated that less fertilizer would result in progressive depletion of the organic nitrogen of the soil. New calculations in 1975 led to increased recommendations of 100 kg per hectare, and between 1975 and 1985 farmers increased their fertilizer usage steadily to 190 kg per hectare. A proportion of the nitrogen was taken up by the crop, but any excess would be available for leaching from the soil and into the groundwater.

The rate of leaching from the soil depends on a wide variety of factors. Nitrate, like all soluble ions, moves with the soil water and, under plant cover, root uptake combined with evaporation of water and the temperature gradient between soil and air results in a general upward movement of nitrate. In contrast, in a vegetation-free period and high rainfall there is a general downward movement of the nitrate into the aquifers. In general, as the amount of water draining from the top soil to the deeper layers increases, so does the rate of nitrate leaching from the soil (Amberger 1974).

3.5 NITRATE LEACHING FROM THE SOIL

(i) Plant cover

Although nitrate leaches from all soils regardless of the extent of plant cover or fertilizer usage (Furrer 1986), bare soils lose greater amounts of nitrate through leaching because of the lack of uptake by plant roots (Cooke 1984). The presence or absence of plant cover thus has a much greater influence on the rate of nitrate leaching than any other factor including the rate of fertilizer application (Furrer 1986), as illustrated in Table 3.5, which shows that leaching from cropped soils is much less than from bare soils. Table 3.6 shows the fate of nitrate fertilizer when applied to soil with good crop cover.

The type of plant cover is influential; the effectiveness in inhibiting leaching decreases from woodland to grassland to arable land (Atkins 1976, Cook, 1984). When active growth of the plant cover ceases, the active uptake of nitrate also ceases, as does the effect on nitrate leaching.

Table 3.6 — The fate of administered fertilizer nitrogen applied to crops

Taken up by crop plants	50–75%
Bound as organic-N in the soil	5–25%
Bacterial breakdown to N_2 gas	2–20%
Leached into the subsoil	2–10%

(ii) Effect of rainfall or irrigation

The important factor here is the rate of rainfall in comparison to the rate of evaporation of water from the soil. Thus autumn and winter rainfall is of more importance because it occurs when the temperature is cooler and the plant cover is less.

In contrast, irrigation usually occurs during the summer and, by encouraging maximal crop growth rate, also encourages maximal uptake of nitrate and so minimal leaching into the groundwater (Webster *et al.*, 1984). Excessive irrigation, of course, will have a counter-effect and increase nitrate leaching. To the extent that irrigation increases the total flow of water through the soil, there may be a net increase in nitrate loss from the soil (ECETOC 1988).

(iii) Effect of temperature

Temperature affects nitrate leaching by a number of mechanisms. As the temperature increases so the rate of evaporation from the soil also increases and the rate of leaching slows. In the presence of crop cover, the rate of growth of the crops increases (as does the rate of nitrate uptake) and so increased temperature is accompanied by a transfer of nitrate from the water-soluble to the organic phase and this, again, slows the rate of leaching. There will also, however, be an increase in the rate of nitrate formation by soil organisms and, in the absence of crop cover, to utilize it, this will tend to result in increased nitrate leaching.

(iv) Type of soil

The major soil characteristics of relevance are the depth of soil, the texture and structure, and the parent material. Shallow soils are usually well drained and so are associated with rapid leaching. Soils rich in organic matter tend to be rich in organic nitrogen and so provide a suitable material for bacterial formation of nitrate and subsequent nitrate leaching (Allison 1966, Royal Society Study Group 1983). Sandy and light soils tend to retain water less well than clay/heavy soils and so allow more free movement of water (and soluble ions) following, for example, rainfall. In contrast, clay soils tend to retain water (and nitrate) much better.

Although light sandy soils lose nitrate more readily, the effect of this on the peak nitrate concentration leaching from it tends to be lower. In consequence a high rainfall following the harvesting of crops on light sandy soil does not tend to give a high peak nitrate leaching because relatively little is retained. In contrast, a heavy clay soil tends to retain its nitrate load so that, when drainage begins in the autumn, the peak nitrate level in the groundwater can achieve high values. It is the peak value that is likely to be of greater importance with respect to compliance with legislation.

(v) Groundwater level and drainage

Soils with poor natural drainage, a large water-holding capacity or a low water table all tend to have lower nitrate losses through leaching because the increased residence time of the water in the soil results in increased uptake of the nitrate by plants. Artificial drainage, for the same reasons, increases nitrate leaching (Colbourn 1985).

(vi) Crop strains

Improvements in strains of crops have resulted, amongst other things, in improved nitrate utilization and greater depletion of soil nitrate (Bommer 1982, Zerulla 1986). Jenkinson (1982) has reported that, in a trial situation, whereas wheat used only 32% of the fertilizer supply in 1967, by 1980 this had increased to 86%.

(vii) Form of nitrogen

The form of low-molecular-weight water-soluble nitrogen has little effect on the rate of leaching, since non-nitrate nitrogen is rapidly converted to nitrate in the soil.

3.6 QUANTITATIVE ASPECTS

All estimates of the relative proportions contributed by the various sources are, of course, subject to considerable regional variation. Thus in sources that are close to major conurbations the contribution from sewage may be particularly high. However, probably the most contentious issue is the contribution of agricultural fertilizer.

On the model developed by the Water Research Centre and used by the UK Department of the Environment in its 1988 analysis of 'the nitrate problem', nitrogen equivalent to 40% of the amount applied as fertilizer leaks into the groundwater as nitrate. This has been challenged by Addiscott and Powlson (1989) on the basis of work carried out at the AFRC Experimental Station at Rothamsted. Work at Rothamsted showed that even soil that had been treated with no fertilizer could leak nitrate at the rate of 20 kg per hectare per year; Goss *et al.*, at Rothamsted, showed no relation between the nitrate in water draining from various plots and the amount of nitrogen fertilizer applied *during the spring.* Addiscott and Powlson claim that little or no nitrate fertilizer leaches from the soil into the groundwater because there is very little not taken up by, for example, winter wheat. However, as the amount of nitrogen fertilizer was increased, the amount of nitrate from other sources leaching into the groundwater increased.

Table 3.7 — Factors that maximize nitrate leaching into soil

Fertilizer	Autumn application (when ground cover is less)
	Spring sowing (so reducing winter ground cover)
Grassland	Ploughing releases nitrate over a prolonged period
	Age grassland increases NO_3
	Minimal ground cover
	Added nitrate fertilizer speeds break-up
Animal manure	Rapid breakdown to nitrate by bacteria

A study of the origins of groundwater nitrate began in Rothamsted in 1870. Soils that received no fertilizer and which grew no crops nevertheless yielded nitrate into the groundwater, draining the soil at the rate of 45 kg nitrate per hectare per year. Of this, only 5–10% was from the rainwater and the remainder matched the organic nitrogen lost from the soil.

Table 3.7 illustrates and summarizes the factors that maximize the level in drinking water and groundwater.

REFERENCES

Addiscott, T. (1988). Farmers, fertilizers and the nitrate flood. *New Scientist* (8 Oct), 50–54.

Addiscott, T. and Powlson, D. (1989). Laying the ground rules for nitrate. *New Scientist* (29 April), 28–29.

Allison, F. E. (1966). The fate of nitrogen applied to soils. *Adv. Agron.* **18**, 219–258.

Anderson, K. K. and Hooper, A. B. (1983). O_2 and H_2O are each a source of one 'O' in NO_2 produced from NH_3 by *Nitrosomonas*. *FEBS Lett.* **164**, 236–240.

Atkins, S. F. (1976). Nitrogen leaching from fertilizers: lysimeter trials. Published results from Europe and USA. *ICI Central File No. A128*, **607**, 1–6 and appendix.

Beijerinck, M. W. (1888). Cultur der *Bacillus radicola* aus den Knollchen. *Bot. Ztg* **46**, 740–750.

Beijerinck, M. W. (1901). Ueber oligonitrophile Mikroben. *Zbl. Bakt. Itc. Abt.* **7**, 561–582.

Beijerinck, M. W. and van Delden, A. (1902). Uber die assimilation desfrcien Stickstoffs durch bakterien. *Zbl. Bakt.* **9**, 3–43.

Bewley, W. F. and Hutchinson, H. B. (1920). On the changes through which the nodule organism (PS Radicicola) passes under cultural conditions. *J. Agric. Science* **10**, 144–162.

Bommer, D. F. R. (1982). Landwirtschaft zwischen mangel und uberfluss — analysen und perspektiven. *BASF-Mitt Landbau* **5**, 1–33.

Bishop, P. E., Jarlenski, D. M. and Hetherington, D. R. (1980). Evidence of an alternative nitrogen fixation system in *Azotobacter vinelandii*. *PNAS* **77**, 7342–7346.

Christy, M., Brown, J. R. and Smith, G. (1973). Nitrate and water. Univ. of Missouri — *Columbia Sci. and Tech. Guide.*

Colbourn, P. (1985). Fertiliser nitrogen: where does it all go? Paper presented at National Agric. Conference *Better Usage of Nitrogen — the Prospects for Winter Cereals*, London Jan 23.

Cole, J. A. (1988). Assimilatory and dissimilatory reduction of nitrate to ammonia. In: *The Nitrogen and Sulphur Cycles.* Eds J. Cole and S. Ferguson, Cambridge University Press, pp.281–330.

Cooke, G. W. (1984). Fertiliser application versus crop requirements. Presented at EC-Symposium *Environment and Chemicals in Agriculture*, Dublin, pp. 15–18, October.

Cutler, D. W. and Mukerji, B. J. (1931). Nitrite formation by soil bacteria other than Nitrosomonas. *Proc. Roy. Soc. B.* **108**, 384–394.

Davey, K. W. (1970). An investigation into the nitrate pollution of the chalk borehole water supplies. North Lindsey Water Board.

Department of the Environment (1988). *The Nitrate Issue.* HMSO, London.

Dispirito, A., Taaffe, L. and Hooper, A. B. (1985). Localisation and concentration of hydroxylamine oxidoreductase and cytochromes C-552, C554, C_m0552 and a in *Nitrosomonas europaea. Biochem. Biophys. Acta.* **806**, 320–330.

Dutchion, C. (1985). Landspreading of pig manure. 1. Effects on yield and quality of crops. In: *Effluents from Livestock.* Ed J. Grasser. Applied Science Publishers, London, p. 32–58.

Eady, R. R. (1986). Enzymology in free-living diazotrophs. In: *Nitrogen Fixation: Vol 4.* Eds W. Broughton and A. Puhler, Clarendon Press, Oxford, pp. 1–49.

Eady, R. R., Robson, R. L. and Smith, B. E. (1988). Alternative and conventional nitrogenases. In: *The Nitrogen and Sulphur Cycles,* Eds J. Cole and S. Ferguson, Cambridge University Press, Cambridge, pp. 363–380.

ECETOC (1988). *Nitrate and Drinking Water.* ECETOC Technical Report No. 27.

Fliermans, C. B., Bohlool, B. B. and Schmidt, E. L. (1974). Autoecological study of the chemoautotroph Nitrobacter by immunofluorescence. *Appl. Microbiol.* **27**, 124–129.

Forget, P. (1971). Les nitrate-reductases bacteriennes. Solubilisation, purification et proprietes de l'enzyme Ade *Micrococcus denitrificans. Eur. J. Biochem.* **18**, 442–450.

Foster, S. S. D., Cripps, A. C. and Smith-Carrington, A. (1982). Nitrate leaching to groundwater. *Phil. Trans. Roy. Soc. (London)* **296**, 477–489.

Fred, E. B. and Davenport, A. (1918). Influence of reaction on nitrate-assimilating bacteria. *J. Agric. Res.* **14**, 317–336.

Furrer, O. J. (1986). Einfluss Fruchtfolge und Dungung auf den Nitrateintag ins Grundwasser. *10th Int. Fachmesse und Fachtagungen fur Umwelt schutztechnik und Okologie.* Basel.

Garcia, J. L. (1973). Sequence des produits formes au cours de la denitrification dans les sols de rizieres du Senegal. *Ann. Microbiol. (Paris)* **124B**, 351–362.

Hollocher, T. C. (1984). Source of the oxygen atoms of nitrate in the oxidation of nitrite by *Nitrobacter agilis* and evidence against a P–O–N anhydride mechanism in oxidative phosphorylation. *Arch. Biochem. Biophys.* **233**, 721–727.

Hollocher, T. C., Tate, M. E. and Nicholas, D. J. (1981). Oxidation of ammonia by *Nitrosomonas europaea.* Definitive ^{18}O-tracer evidence that hydroxylamine formation involves a monoxygenase. *J. Biol. Chem.* **256**, 10834–10836.

Holmes, P. A. C. (1974). Investigation into the nitrate pollution of water supply.

Hooper, A. B. and Terry, K. R. (1973). Specific inhibitors of ammonia oxidation in *Nitrosomonas. J. Bacteriol.* **115**, 480–485.

Hooper, A. B., Maxwell, P. C. and Terry, K. R. (1978). Hydroxylamine oxidoreductase from *Nitrosomonas*: absorption spectra and content of heme and metal. *Biochemistry* **17**, 2984–2989.

Jenkinson, D. S. (1982). The nitrogen cycle in long term field experiments. *Phil. Trans. Roy. Soc.* (London) **B296**, 563–571.

Keele, B. B., Hamilton, P. B. and Elkan G (1969). Glucose catabolism in *Rhizobium japonicum*. *J. Bacteriol.* **97**, 1184–1191.

Killham, K. (1986). Heterotrophic nitrification. In: *Nitrification*. Ed. J. Prosser, IRL Press, Lancaster, pp. 117–126.

Krummel, A. and Harms, H. (1982). Effect of organic matter on growth and cell yield of ammonia-oxidising bacteria. *Arch. Microbiol.* **133**, 50–54.

Kumar, S., Nicholas, D. and Williams, E. (1983). Definitive ^{15}N-NMR evidence that water serves as a source of 'O' during nitrite oxidation by *Nitrobacter agilis*. *FEBS Lett.* **152**, 71–74.

Orme-Johnson, W. H. (1985). Molecular basis of biological nitrogen fixation. *Ann. Rev. Biophys. Chem.* **14**, 419–459.

Payne, W. J. (1973). Reduction of nitrogenous oxides by microorganisms. *Bact. Rev.* **37**, 409–452.

Payne, W. J. and Riley, P. S. (1969). Suppression by nitrite of enzymatic reduction of nitric oxide. *Proc. Soc. Exp. Biol. Med.* **132**, 258–260.

Payne, W. J., Riley, P. S. and Cox, C. D. (1971). Separate nitrite, nitric oxide and nitrous oxide reducing fractions from *Ps. perfectomarinus*. *J. Bacteriol.* **106**, 356–361.

Payne, W. J. and Balderston, W. L. (1978). Denitrification. In: *Microbiology-1978*. Ed. D. Schlessinger. *Am. Soc. Microbiol.*, Washington, pp. 339–342.

Pedrosa, F., Dobereiner, J. and Yates, M. (1980). Hydrogen dependent growth and autotrophic carbon dioxide fixation in *Derxia*. *J. Gen. Microbiol.* **119**, 547–551.

Pichinoty, F. (1973). La reduction bacterienne des composes oxygenes mineraux de l'azote. *Bull. Inst. Pasteur. (Paris)* **71**, 317–395.

Ronson, C. W. and Primrose, S. B. (1979). Carbohydrate metabolism in *Rhizobium trifolie*. Identifiction and symbiotic properties of mutants. *J. Gen. Microbiol.* **112**, 77–88.

Royal Society Study Group (1983). *The Nitrogen Cycle in the UK*. Royal Society Report, London, pp. 1–264.

Ryden, J. C., Ball, P. R. and Garwood, E. A. (1984). Nitrate leaching from grassland. *Nature* **311**, 50–53.

Sims, C. M. and Collins, F. M. (1959). Nitrite production by athermophilic bacterium. *Austr. J. Agric. Res.* **10**, 832–838.

Suzuki, I. (1974). Mechanisms of inorganic oxidation and energy coupling. *Ann. Rev. Microbiol.* **28**, 85–101.

Tanaka, Y., Fukumori, Y. and Yamanaka, T. (1983). Purification of cytochromes a_1 c_1 from *Nitrobacter agilis* and characterisation of nitrite oxidation system of the bacterium. *Arch. Microbiol.* **135**, 265–271.

Thompson, J. P. and Skerman, V. B. D. (1979). *Azotobacteraceae: The Taxonomy* and Ecology of Aerobic Nitrogen-fixing Bacteria. Academic Press, London.

Tudge, C. (1984). Whatever happens to nitrogen? *New Scientist* (9 Feb) 13–15.

Van Dijk, T. A. and Sturm, H. (1983). Fertilizer value of animal manure on the continent. *Proc. Fert. Soc. (London)* **220**, 1–45.

Vetter, H. and Steffens, G. (1983). Stickstoffaustrag aus vorweigend organisch gedungten Flachen insbesondere mit Gulle. In 'Nitrat: Ein Problem fur unsere Trinkwasserversorgung?' *Arbeiten der DLG* **177**, 121–133.

Walker, N. (1978). On the diversity of nitrifiers in nature. In: *Microbiology-1978.* Ed. D. Schlessinger, *Am. Soc. Microbiol.*, Washington, pp. 346–347.

Webster, C. P., Belford, R. K. and Cannell, R. Q. (1984). Crop uptake and leaching losses of ^{15}N labelled fertilizer nitrogen in relation to waterlogging of clay and sandy loam soils. *Plants and Soils* **92**, 89–101.

WHO (1985). *Health Hazards for Nitrate in Drinking Water.*

Winogradsky, S. (1890). Recherches sur les organismes de la nitrification. *Ann. Inst. Pasteur (Paris)* **4**, 213–231.

Winogradsky, S. (1891). Recherches sur les organismes de la nitrification. *Ann. Inst. Pasteur (Paris)*, **5**, 577–647.

Winogradsky, S. and Winogradsky, H. (1933). Nouvilles recherches sur les organismes de la nitrification. *Ann. Inst. Pasteur (Paris)*, **50**, 350–432.

Yates, M. G. (1988). The role of oxygen and hydrogen in nitrogen fixation. In: *The Nitrogen and Sulphur Cycles.* Eds J. Cole and S. Ferguson, Cambridge University Press, pp. 383–416.

Zerulla, W. (1986). Die Leistung des Stickstoffs un Wanderder Zeiten un Abhangigkeit von verschiedenen Produktionssystemen — darstellt am Beispeil Getreide. *BASF Mitt. Landbau* **HI**, pp. 1–62.

4

Origin and significance of nitrite in water

C. W. Mackerness and C. W. Keevil

4.1 INTRODUCTION

Nitrite and nitrate play an important role in the nitrogen cycle (Walters, Chapter 5) involving the complex interaction of different ecosystems of the biosphere. Both nitrate and nitrite are present in water and food, and it is by these sources that humans are exposed to these ions. Therefore, ingested food and water are the main sources of nitrate and nitrite exposure in an individual, with the other sources (for example the respiratory burst of macrophages) (Stuehr and Marletta 1987) thought to have only a marginal influence. There is concern in the UK over the increasing concentrations of nitrate and nitrite in drinking water in some areas. Recent legislation, EC Directive 80/778/EEC, has set the maximum permissible concentrations of nitrate at 50 mg/l and nitrite at 0.1 mg/l in water at the point of use. In this chapter we shall examine the origin of nitrite in the domestic drinking water supply, with regard to the possible effects of this ion on human health.

Although the methods of nitrite determination have varied over the years (Massey, Chapter 1) estimations of nitrite in water have shown that, when compared to nitrate, the concentration of nitrite is generally very low. The limits set in the EC Directive reflect this. However, some water utilities have experienced problems of non-compliance with the EC Directive for the maximum permissible concentration of nitrite in potable drinking water, which are higher than the concentrations in the source (raw) water. The mechanism by which nitrite is formed in the distribution system has recently been elucidated (Mackerness *et al.* 1990).

There are various suggested health effects resulting from the ingestion of high concentrations of nitrite, but many studies describe experiments on animals (usually rats) where unrealistic quantities of ingested nitrite were used: hence difficulties arise in deducing the toxicological effects of nitrite in man. Fortunately there have been other studies where, for example, accidental exposure of humans gives more reliable evidence of deleterious effects of ingested nitrite. While the acute toxic effects of large doses of nitrite are well documented, the significance of relatively small concentrations of nitrite to adults is judged mainly in the light of the possible role of nitrite as a precursor of the carcinogenic N-nitroso compounds (NNC). For this reason it is important to identify those factors thought likely to increase nitrite ingestion in order to reduce the intake of this ion since there is evidence that high intra-gastric nitrite concentration correlates with an increased risk from stomach cancer (Hartman 1982). The reduction of nitrate to nitrite in the gastric lumen is an important source of nitrite for the formation of NNC. It has been suspected for some years that the effect of exposure to nitrate and nitrite may cause human cancer (Mirvish 1983); however, the epidemiological evidence relating nitrate exposure to cancer has so far been inconclusive. There are problems in attempting to test the hypothesis to prove a direct relationship since gastric cancer has a multi-factorial aetiology.

4.2 THE ORIGIN OF NITRITE IN WATER

The ingestion of nitrite is low, when compared to nitrate ingestion, and certain foods, such as cured meats and some root vegetables, contribute the greatest amount of nitrite, with the drinking water supply contributing only a very small amount. In the UK, the average estimated intake of total dietary nitrite lies between 0.3 and 0.9 mg per person per day, with root vegetables making the greatest contribution to the estimated intake, with the mean nitrite intake in food estimated to be 1.4 mg (Knight *et al.* 1987). However, it has been estimated that in the US population, the average daily dietary intake of nitrite ranges from 0.01 to 2.6 mg per person per day (Fan *et al.* 1987). Hence there is considerable variation between populations in the estimates of dietary nitrite intake; some reported values are shown in Table 4.1.

Table 4.1 — The mean per capita daily nitrite ingestion (mole/person/day) of various countries

Country		Reference
W. Germany	0.022	Sander (1971)
Norway	0.0024	Höyem (1974)
US	0.054	White (1975, 1976)
Sweden	0.08 (0.013–0.24)	Jägerstad and Nordén (1976)
W. Germany	0.04	Selenka and Brand-Grimm (1978)
W. Germany	0.07 (0.05–0.08)	Selenka and Brand-Grimm (1978)
Netherlands	0.097	Stephany and Schuller (1978)

Further, in the US the human stomach is estimated to receive 4–6 mg NO_2 per day (NAS 1981), with the estimated fasting concentration of 120 µg/l nitrite in the stomach.

The contribution of nitrite in water as a source of dietary nitrite is considered to be negligible since generally nitrite is not detected in drinking water leaving the treatment works. Water of the distribution system sometimes contains concentrations of nitrite in the range 0.1–0.3 mg NO_2^- (MAFF Surveillance paper No. 20). If this situation is the case, and if the average person drinks one litre of water per day, then the maximum intake of nitrite from drinking water might be 0.2 mg per person per day from this source. Higher nitrite concentrations have been found in association with bacterial contamination of water, but such water would in any case be considered unfit for human consumption on microbiological grounds. There have been various estimates of drinking water nitrite in different countries: it has been reported (Alwan *et al.* 1987) that the concentration of nitrite in water in Baghdad, Iraq, was in the range 26–67 µg/l (± 20), whereas Selenka and Brand (1975) estimated that the concentration of nitrite in the drinking water of the Rhineland Palatinate (West Germany) was 1–2 mg/l and concluded that this was therefore unlikely to be of any public health significance.

The formation of nitrite in the drinking water supply can be from two potential mechanisms, either chemical or microbiological. The chemical formation of nitrite is by either the oxidation of ammonia, or the reduction of nitrate:

$$NH_3 \underset{\text{reduction}}{\overset{\text{oxidation}}{\rightleftharpoons}} NO_2 \underset{\text{reduction}}{\overset{\text{oxidation}}{\rightleftharpoons}} NO_3$$

In addition there are occasions where ammonia is present in the water, either naturally or added deliberately during treatment to convert free residual chlorine to chloramine. There is no evidence to suggest that in practice nitrite in water is the result of purely chemical processes. Further, there are too few bacteria in the water to account for the nitrite production. Thus, if nitrite is found at high concentrations in mains water its appearance must be due to the action of bacteria in surface biofilms. Bacteria and fungi are known to be able to colonize and form biofilms of mixed microbial populations acting as consortia on all of the materials used in the water distribution system (Keevil *et al.* 1989). The precise detail of the complex microbial consortium in the biofilm, or indeed the microflora's metabolic capabilities, have not yet been properly examined.

Until recently, little attention has been paid to these organisms, nonetheless they act as a focus of metabolic activity in distribution systems and have the capacity to alter drinking water quality. In general, water utilities monitor distribution system water only for the presence of coliforms as an indicator of faecal contamination, and do not consider the other non-pathogenic organisms which are also found in the supply as being a possible cause of water-borne enteric infections. Therefore, since the purpose of disinfection is only to eliminate pathogenic organisms, potable water is not sterile.

In addition, the biofilm may be able to protect coliform bacteria from the disinfectants currently in use (Keevil *et al.* 1990, Mackerness *et al.* 1990). The sloughing-off or release of coliforms from the biofilm could lead to their detection in potable water (treated to WHO and EC guidelines) where there is no known evidence of faecal contamination. The bacteria sloughed off from the biofilm, though non-pathogenic, may nevertheless be capable of colonizing the stomachs of achlorhydrics, and these organisms may reduce nitrate to nitrite or catalyse N-nitrosation reactions in the stomachs of these individuals.

Since there is also little current information available on the possible mechanisms of nitrate or nitrite formation in water distribution systems, we have attempted to determine those factors which may be important in nitrite formation in drinking water supply systems, utilizing laboratory-based experiments which model the environmental conditions of individual supplies (Mackerness *et al.* 1990). In the model, ammonia was stoichiometrically converted to nitrite after an initial lag of seven days. It is, therefore, important to establish whether this is also the origin of nitrite in drinking water supply systems in order to devise strategies that may lower its concentration in drinking water.

Nitrite formation was detected only recently in London's distribution system as a result of the change of emphasis in the early 1980s on monitoring from source works to the point of use. The mechanism of nitrite formation was unclear, but empirical studies showed a link with the residual disinfectant, monochloramine. For many years, river-derived water in London has been disinfected at the treatment works, using free chlorine. Natural ammonia in the source water reacts with chlorine, which is added in excess to generate sufficient free chlorine to act as the primary disinfectant (breakpoint chlorination). After the disinfection contact time the concentration of free chlorine is reduced by the addition of sulphur dioxide and then converted to monochloramine by the deliberate addition of ammonia. Water thus enters the distribution system with a controlled amount of monochloramine and excess ammonia. The choice of monochloramine as the residual disinfectant was based on the need for penetration of the residual chlorine throughout the long and complex distribution system of service reservoirs and mains.

As there have been occasional problems associated with certain areas of London's distribution system with regard to nitrite non-compliance to the EC regulations, these supply works have been investigated microbiologically. Since the process of post-ammoniation is suspected as being a contributing factor in the formation of nitrite, particular emphasis has been placed on determining those organisms capable of oxidizing ammonia to nitrite in the distribution supply. Some of the bacteria in treated water supplies (i.e. the nitrifying organisms) are able to oxidize ammonia to nitrite (for example *Nitrosomonas* spp) or oxidize nitrite to nitrate (for example *Nitrobacter* spp). Those bacteria which are capable of oxidizing ammonia to nitrite and/or nitrate can be broadly classified into two groups of organisms: the autotrophic nitrifiers and the heterotrophic nitrifiers.

Nitrification, catalysed by bacteria, includes two major steps: the first is the oxidation of ammonia to nitrite by *Nitrosomonas* spp whilst the second is the oxidation of nitrite to nitrate by *Nitrobacter* spp. *Nitrosomonas* spp obtain energy by the oxidation of ammonia to nitrite, and use CO_2 as the carbon source. They are rods, often motile with polar flagella. The optimum growth temperature is between 25 and 30°C, and the optimum pH between 7.5 and 8.0, and require aerated conditions. The mechanism of the oxidation of ammonia was proposed by Suzuki (1974):

(1) $NH_3 + O_2 + AH_2 ? NH_2OH + H_2O + A$
(2) $NH_2OH + 2cyt.c\ (Fe^{3+}) ? NOH + 2H^+ + 2cyt.c\ (Fe^{2+})$
(3) $2cyt.c\ (Fe^{2+}) + \frac{1}{2}O_2 + 2H^+ ? 2cyt.c\ (Fe^{3+}) + H_2O$
(4) $NOH + A + H_2O ? NO_2 + AH_2 + H^+$

In reaction (1), O_2 is molecular oxygen, and in reaction (4), O_2 is from water. The oxidation of hydroxylamine to nitrite (reactions (2)–(4)) is performed by *Nitrosomonas* and *Nitrosococcus*: the enzyme responsible for reaction (1) is ammonia oxidase, whilst the enzymes of reactions (2)–(4) are termed the hydroxylamine oxidoreductase; in fact there are two enzymes of this complex, in reaction (2), hydroxylamine dehydrogenase, and reaction (4), nitrite synthase.

In addition, facultative bacteria such as faecal pathogens can also reduce nitrate to nitrite anaerobically. Metabolically active biofilms in the distribution systems may develop anoxic regions, permitting anaerobic activity and hence survival of oxygen-sensitive bacteria which may be of significance to public health. Thus metabolic reactions which aid survival of these microorganisms *in vitro*, and affect the steady-state conditions of ammonia, nitrite and nitrate (nitrogen balance), may also occur *in vivo*, altering the nitrite concentration and possibly affecting NNC formation.

The autotrophic nitrifiers are Gram-negative, chemolithotrophic, fix carbon via the Calvin cycle for cellular carbon, and derive energy by the reducing power of the oxidation of ammonia to nitrite. Organisms with these characteristics belong to the Nitrobacteraceae family, the genera of which oxidize ammonia to nitrite are: *Nitrosomonas*, *Nitrosococcus*, *Nitrosovibrio*, *Nitrosospira*, and *Nitrosolobus*; the most commonly occurring group are the *Nitrosomonas*. The oxidation of nitrite to nitrate is carried out by the chemoautotrophs of the genera *Nitrosococcus*, *Nitrobacter*, and *Nitrospira*.

The first metabolic step in the oxidation of ammonia is the cytochrome-dependent hydroxylation of ammonia to hydroxylamine, which is an energy-requiring process. The oxidation of hydroxylamine through to nitrite is coupled to ATP generation via a membrane-bound transport system. The terminal electron acceptor is molecular oxygen, hence the process of nitrification is aerobic. Although some species of *Nitrobacter* are capable of autotrophy and heterotrophy the metabolism of nitrifying bacteria is predominantly autotrophic. Under anaerobic conditions some *Nitrobacter* species have been shown to reduce nitrate to nitrite, and further to ammonia and nitrogen gases, particularly nitrous oxide (N_2O). Thus nitrite formation may be particularly susceptible to the local micro-environment redox potential in the planktonic phase and the sessile phases of the distribution system pipework, and the stomach wall and lumen.

Heterotrophic nitrification is the process of ammonia oxidation to nitrite, and differs from autotrophic nitrification in that nitrification is not linked to cell growth. Although intermediates such as hydroxylamine are produced during the growth phase, major nitrification products are formed at stationary phase. The nitrification rates of heterotrophs are in the order of 10^3–10^4 times less than autotrophs. Although heterotrophic nitrification is well documented in laboratory culture, it is thought to be of little significance in natural systems. However, heterotrophic nitrification produces important intermediates, such as the hydroxamic acids. It has been shown that hydroxamic acids are involved in the uptake of iron by *Arthrobacter* spp. In iron, restricted systems such as neutral pH waters where Fe^{3+} is insoluble, and *in vivo* where iron is tightly complexed with transferrin and lactoferrin siderophores, this ability to 'liberate' essential iron may account for the survival of metabolically poor heterotrophic nitrifiers in competitive microbial populations. Conversely, in iron-rich systems, very little hydroxamate is produced, and nitrite and hydroxylamine concentrations increase.

Nitrification can potentially have either beneficial or adverse effects on drinking water quality, depending upon the equilibrium position of the reactions. Rittman and Snoeyink (1984) observed that the conversion of ammonia to nitrate resulted in a stable water that did not promote the growth of bacteria in the distribution system, and this practice has been adopted by several European water utilities. Early reports (Larson 1939) showed that incomplete nitrification, i.e. the formation of nitrite, in a distribution system containing approximately 1 mg/l chloramine caused the loss of disinfectant residual, and an increase in heterotrophic plate counts. Similar problems have been documented in USA water utilities that have recently changed from a free residual to a combined monochloramine residual because of high trihalomethane concentrations (Woolfe *et al.* 1988). In a study of two covered reservoirs in two different Californian treatment works, severe nitrification problems were encountered following the change from free to combined residual. Nitrification leads to a rapid decline in chlorine residual, elevated nitrite concentration, and increases in the heterotrophic organisms.

Nitrifying bacteria have been shown to be resistant to disinfectants. For example, nitrifiers proliferated in a reservoir containing a chlorine residual of 1.4–1.5 mg/l. Feben (1935) observed that ammonia-oxidizing bacteria could survive 2 mg/l chlorine for 60 min., whereas *Escherichia coli* was completely inactivated after 2.5 min. It was important in these studies that the cells were washed of nitrite since associated nitrite could considerably reduce the free chlorine concentration. It has been suggested (Winogradsky 1949) that nitrifiers produce a capsular layer and grow in clumps (a biofilm?) that afford protection against disinfectants and toxic compounds. The attachment of nitrifiers to solid surfaces enhances their growth, and may render them more resistant to toxic compounds.

Little is known of the occurrence of nitrifiers in the distribution system, due partly to the difficulty in the isolation and enumeration of these organisms from environmental samples. Nitrifiers can be enumerated in drinking water by the most probable number (MPN) method. Although it has been noted that this method may underestimate nitrifiers (Belser and Mays 1982) it is useful if nitrification rates are simultaneously determined. A possible explanation of the underestimation by MPN may be due to the fact that not all genera and strains are able to grow on any one medium.

As previously stated, the microbiological reduction of nitrate is a further possible source of nitrite in the drinking water supply. There is a diverse range of bacteria capable of nitrate reduction, including aerobes, facultative anaerobes and strict anaerobes. In these organisms, nitrate is used as the terminal electron acceptor instead of molecular oxygen, hence nitrate dissimilation occurs in low-oxygen-level environments: these might include some biofilms. The bacterial nitrate reductase is a term used to describe at least two distinct enzymes that reduce nitrate to nitrite using intracellularly derived electron donors. These reductive processes have two distinct functions to the organism. However, enzymes within these two groups are heterogeneous in terms of their precise structure and electron donor organization.

The first type of nitrate reductase is assimilatory and functions to provide the organism with a nitrogen source, i.e. ammonia, which can be incorporated into amino acids. This enzyme is produced irrespective of oxygen availability, and its expression is controlled by exogenous ammonia and amino acids.

The second type of nitrate reductase is dissimilatory, and can be expressed either for nitrate respiration or for denitrification. Denitrifiers are non-fermentative facultative bacteria that derive energy by nitrate respiration. Non-denitrifiers are aerobes, facultative anaerobes, and strict anaerobes which comprise a more restricted group compared to the denitrifiers and includes members of the *Enterobacteriaceae* and some *Clostridia* spp.

The nitrate dissimilators can be split into two groups: the denitrifiers dissimilate nitrate to gaseous oxides of nitrogen, nitric oxide and nitrous oxide, or to molecular nitrogen, whilst the non-denitrifiers reduce nitrate to ammonia. Those bacteria capable of denitrification (i.e. form gaseous nitrogen from either nitrate or nitrite) are thus potentially able to change nitrate and nitrite concentration in drinking water.

If nitrate dissimilation was a significant cause of the increase of nitrite in drinking water, one would expect to observe a correlation between nitrate and nitrite concentrations. Although this association has not been demonstrated, nitrate dissimilation may be important in nitrite accumulation at those parts of the distribution supply where the flow rate is slow and oxygen concentrations low, which would then tend to favour these anaerobic processes.

The presence of a biofilm in the distribution system may therefore alter the quality of the water, and clearly this indigenous flora warrants further proper investigation. There are operational difficulties in the examination of distribution-supply biofilm (since this would interrupt supply to customers, and could potentially further contaminate the water with soil microorganisms), and in ascribing chemical effects to the activity of the biofilm in a water supply network. It is therefore necessary to devise a method of investigating biofilm formation and metabolism in a laboratory model (Keevil *et al.* 1988), which reflects conditions in the active system. In our laboratory, biofilms have been studied using a novel application of the technique of continuous culture. Surfaces can be introduced into the chemostat, organisms are able to grow on these surfaces and, more importantly, can be aseptically recovered as viable organisms on the appropriate medium. This approach, to suspend a surface into the culture liquid for the growth of biofilm, can model the growth of organisms on the surface of the distribution-supply network. There has been a lot of recent interest in biofilms in a variety of environments, with excellent reviews by Costerton *et al.* (1987) and van Loosdrecht *et al.* (1990), and it is now clear that biofilms play a very important role in natural aquatic and industrial aquatic environments, medical biomaterials, and biodeterioration. Biofilms can be thought of as consortia of organisms of different species along with their extracellular products which are irreversibly attached to a surface but which can also slough off by ill-defined mechanisms. The formation of a biofilm may be an important survival strategy of organisms, since biofilm bacteria have a greater metabolic activity than planktonic organisms, and can protect organisms from unfavourable conditions (Ellwood *et al.* 1982).

The formation of a biofilm in the water supply network has been recognized as being an important factor in water quality (Keevil *et al.* 1989). Although various methods are used to disinfect drinking water, these have been evaluated empirically in the field using bacteria in liquids, not attached to surfaces. The formation of a biofilm seriously compromises the efficacy of the intended biocide (Keevil *et al.* 1990, Le Chevalier *et al.* 1988). The difficulty of assessing a potential biocide in the distribution system makes the use of laboratory models the only satisfactory method of determining the effect of a biocide on biofilm bacteria (Keevil *et al.* 1988). This is particularly true of monochloramine usage by the water industry, discussed previously. The monochloramine concentrations used in potable water appear to be insufficient to penetrate biofilms, or control their metabolic activity (Keevil *et al.* 1990). Thus, if care is not exercised in controlling ammoniation during monochloramine formation, then the ammonia oxidizers of the biofilm can convert the ammonia to nitrite. It is important to note that this metabolic activity does not occur in man, and is therefore a significant process of environmental nitrite exposure.

4.3 THE SIGNIFICANCE OF NITRITE IN WATER

There have been no reported beneficial health effects of nitrite ingested in the diet. Although nitrite toxicity is well known — the lethal dose of nitrite is in the range 1.6–9.5 g (NAS 1981) — a no-effect level has not been established. However, when nitrite and nitrate are present in water, nitrite would have a more pronounced and rapid effect than nitrate, since nitrate has to be reduced first to nitrite by the action of bacteria colonizing the stomach before it can have any toxicological effect. In order to assess the load of nitrite in the average person and hence its impact on man, it is necessary not only to consider the sources but also the fate of nitrite at the particular sites. Appreciable nitrite penetration of the mucosal barrier of the gastro-intestinal tract is thought only to follow erosion and haemorrhage accompanying intense gastritis (Gwatkin and Plummer 1946, Sollman 1957) or intestinal lesions (Finch 1948). Therefore even the very low penetration found after overwhelming doses of nitrite (Naida and Venkatrao 1945) is no indication of events in the physiological situation. It is difficult to accept, then, the assumption that nitrite in the large intestine (if present) finds its way into the bloodstream in anything but minuscule amounts.

It is generally accepted that nitrite, at the usual concentrations found in potable water, does not constitute a hazard to human health. A small proportion of any nitrate contained in ingested water is converted to nitrite, and the extent of this conversion depends upon the subject involved. Relatively reproducible patterns of nitrite formation have been observed in individuals following the consumption of nitrate in water at a concentration equal to the WHO limit (50 mg/l as nitrate) (Walters 1980). The maximum concentration of nitrite was found generally occurring after 1 h. The subjects best able to convert nitrate to nitrite were also the most efficient in using the nitrite once formed, as would be expected by the actions of bacteria with nitrate- and nitrite-reducing properties. On average about 1 mg nitrite is formed per 25 mg nitrate ingested in water by an efficient converter of nitrate to nitrite, assuming a salivary flow of 75 ml/h during an active day. Thus, if one accepts a daily intake of fluid of about 1.5 l (including beverages) with nitrate in water at the concentration of the WHO recommended limit, this would contain 75 mg as nitrate, which would provide 3 mg nitrite on reduction. However, most water supplies contain less nitrate than the WHO recommended limit, so at a typical nitrate level of 10 mg/l as nitrate, 0.6 mg nitrite would be provided daily on reduction.

The nitrite concentration of raw water sources are usually about 1 mg/l as N. However, where the ammonia concentration of drinking and source water is high, there will be elevated levels of nitrite formed from this ammonia by bacterial biofilm metabolism, which could then lead to transient high nitrite concentrations in the stomach.

The study of Walters (1980) found the mean concentration of nitrite in the fasting acid stomach to be 4.9 μM (= 0.22 mg nitrite/l) whilst, in the hypochlorhydric stomach, the normal fasting stomach nitrite concentration was 25.6 μM (= 1.1 mg nitrite/l); therefore the difference between these two values presumably represents the maximum concentration of nitrite which could have reacted under the more favourable conditions of pH. Thus, the highest effective concentration of nitrite to which the normal acid stomach is exposed approximates to 21 μM or 0.9 μM nitrite/l. The mean flow of unstimulated gastric juice (Ruddell *et al.* 1977) was found to be approximately 20 ml/h, which is equivalent to 18 μg nitrite/h or 0.33 mg nitrite in 24 h.

The availability of nitrite in the normal acidic stomach is low. However, in the study of Walters (1980) of 30 patients without demonstrable gastro-duodenal lesions, 12 had a fasting hydrogen ion concentration in the stomach of <5 mM, compared with the mean value of the 18 others of 43.6 ± 5 mM. The mean gastric nitrite concentration of these hypochlorhydric normals was 25.6 ± 3.6 μM compared with 1.7 ± 0.5 μM for normal subjects with fasting pH values of <2.5. A similar inverse relationship, which implies a microbiological association, between nitrite and hydrogen ion concentration was found in cases of disorders of the stomach such as duodenal ulcers and gastric cancer, and thus the effect was due to the resultant pH within the stomach rather than the pathological lesion *per se*.

Concentrations of fasting gastric juice nitrite have been reported to be in the ranges of 0–93 μM normal acidic gastric juice (Leach 1988), and 32–600 θM in achlorhydric gastric juice, and it is this nitrite which could potentially take part in nitrosation reactions to form N-nitroso compounds (NNC). Simply, the process by which NNC are formed, i.e. N-nitrosation, can be defined as the replacement of a hydrogen attached to a nitrogen by a nitroso group. The donor of the nitroso group is termed the nitrosating agent. Mechanisms of N-nitrosation have been recently reviewed by Shuker (1988), Challis and Challis (1982), and earlier by Douglass *et al.* (1978). It is important to note that nitrite *per se* is not a nitrosating agent. There are a number of different nitrosating agents that act as precursors of NNC, *viz*. nitrate, and the oxides of nitrogen (nitrous oxide and nitric oxide). The known nitrosating agents, N_2O_3, N_2O_4, H_2ONO^+ and NO^+, are all derived chemically as a consequence of the behaviour of nitrite in an acid solution, where the balance between the different chemical species being formed depends upon the pH. Of biological significance, it has been shown that intact resting bacteria with functional nitrate or nitrite reductases are able to catalyse NNC formation: so although aquatic bacteria have the potential to generate carcinogens, this is unlikely to occur *in vivo* owing to the nutrient-depleted oligotrophic waters; hence bacteria have to wait for nutrient-rich environments for nitrosating substrates.

It must be noted, however, that the role of nitrite in N-nitrosation reactions in the stomach is complicated by the fact that there are competitive side reactions of nitrite that can occur that can reduce the effective concentrations of nitrite available for reaction with secondary amino compounds and amides. Therefore in the evaluation of nitrite exposure, some consideration must be given to inhibitors and stimulators which are likely to be ingested concurrently or encountered *in vivo*. Hence, the proportion of ingested nitrite which is potentially available for N-nitrosation reactions in the stomach can be affected by other dietary components ingested at the same time, since several workers have shown that *in vivo* and *in vitro* N-nitrosation has been inhibited by ascorbate and, to a lesser degree, α-tocopherol (Mirvish *et al.* 1972, Oshima and Bartsch 1981, Mackerness *et al.* 1989). Vitamin E (α-tocopherol) is a lipid-soluble phenol that reacts with nitrite to produce NO and a quinoid derivative, and can destroy nitrous acid or nitrogen oxides in lipid media. Nitrite (as the dimer N_2O_4) can nitrosate amides and amines in lipid solvents, and vitamin E strongly inhibits these reactions.

For significant quantities of N-nitroso compounds to be produced in the stomach there will be a requirement for the simultaneous presence of nitrite and nitrosatable substrate in the stomach. Gastric pH increases in individuals with mucosal lesions (for example, chronic atrophic gastritis) and also with age (Walters *et al.* 1982). The reduction of gastric acidity allows colonization by bacteria which will include nitrate reducers, and also increases the gastric pH closer to the optimum pH of 6–6.4 for bacterial nitrate reductase (Stouthamer 1976). This increases the nitrite concentration (Reed *et al.* 1981, Ruddell *et al.* 1976), and in individuals with such lesions, intragastric reduction of nitrate may be the most important source of gastric nitrite. Chemical nitrosation by nitrite proceeds maximally at acid pH <5; thus overall N-nitrosation could be decreased at high pH despite the higher nitrite concentration when there is a low level of bacterial nitrosation.

In the achlorhydric stomach and at other body sites at neutral pH, bacterial N-nitrosation reactions at the pH optimum 6–8 are those most likely to be responsible for NNC formation (Leach *et al.* 1990). Under these conditions the acid-catalysed reactions are of little significance. Since the achlorhydric stomach is also probably colonized by nitrate-reducing bacteria there will therefore be elevated levels of nitrite substrate. In spite of the criticisms of the earlier work on bacterial nitrosation (Ralt and Tannenbaum 1981) it has now been conclusively established that certain bacteria are able to catalyse the formation of NNC at neutral pH (Leach 1988, Leach *et al.* 1990).

Nitrite can be formed by the reduction of nitrate *in vivo*, by certain species of bacteria, and such nitrate-derived nitrite can be the main source of exposure to nitrite. Since nitrite ingestion in humans is ordinarily low when compared to nitrate ingestion, it is likely that internal nitrite is derived from ingested nitrate.

The toxicological effects of nitrite are discussed in greater detail by Hill (Chapter 8). There are noteworthy examples of health effects of high nitrite exposure which correlate with high rates of oesophageal cancer in China (Yang 1980) and Iran (Eisenbrand *et al.* 1980). This high nitrite intake of a high-cancer-incidence region of Iran (Joint Iran–International Agency for Research on Cancer Study Group 1977) was ten times that calculated for the US, and furthermore this high nitrite intake was accompanied by low ascorbate ingestion. In animal experiments it has been suggested that nitrite enhanced the carcinogenic effect of N-methylbenzylamine, induced the enhanced progression of oesophageal tumours, and it was also observed that nitrite was responsible for pathological changes of the respiratory system (Schweinsberg and Bürkle 1985).

It is known that nitrite entering or formed in the stomach is adsorbed into the bloodstream, but little is known about the circulation and metabolism of nitrite in the body. It is thought that the nitrite concentration of plasma is negligible (Hartman 1982, Rodkey 1976). The initial reaction between nitrite and oxyhaemoglobin, which is directly dependent upon the concentration of nitrite, is slow until the nitrite salt concentration of methaemoglobin accumulates as a product. Then a faster second reaction occurs which is directly dependent upon oxyhaemoglobin concentration and the square of the concentration of methaemoglobin. During this process the nitrite is converted to nitrate. In some individuals even 'normal' nitrite levels produce acute effects, such as headaches (Henderson and Raskin 1972), whilst in infants they can cause methaemoglobinaemia.

While the acute toxic effects of large doses of nitrite are well documented, the positive data on the developmental toxicity of nitrite are limited. The significance of small amounts of nitrite is mainly judged in the light of the possible role of nitrite as a precursor of the carcinogenic N-nitroso compounds (Schweinsberg and Bürkle 1985).

There have been various studies investigating the toxicity of nitrite in rats. These experiments were concerned mainly with the effect of nitrite on the developing foetus. Shuval and Gruener (1972) found that sodium nitrite in drinking water supplied to pregnant rats caused maternal anaemia and increased infant mortality, whilst Vorhees *et al.* (1984) found that sodium nitrite increased offspring mortality, and at higher doses decreased weight gain. Sodium nitrite fed to pregnant rats at the concentrations of 2 and 3 g/l (Roth *et al.* 1987) resulted in the pups being born anaemic, and in death at the third week *post partum*, although the adult mothers did not suffer any ill effects. Globus and Samual (1978) found that nitrite administered orally at the concentration of 0.5 mg/mouse/day to mice at 10–15 d of pregnancy caused an increase in the asymmetrical fusion of sternebra. Tumours were induced in rats at high nitrite concentrations (Mirvish *et al.* 1980), and the incidence of liver neoplasms significantly increased in female rats given nitrite alone (Lijinsky *et al.*

1983) although under the conditions of several studies (Mackawa *et al.* 1982, Shimada 1989) there was no carcinogenic effect of nitrite. N-nitrosamines were formed from nitrite and amines in one experimental model inducing tumours (Lijinsky 1984).

4.4 CONCLUSIONS

The possibilities of reducing nitrite exposure are not great, so far as nitrite derived from vegetables and water is concerned, since this would be dependent upon reducing their nitrate contents: although exogenous nitrate and nitrite are potentially avoidable sources of intra-gastric nitrite.

The formation of monochloramine in certain treatment works potentially leads to excess ammonia in the distribution system. It is this excess ammonia which is oxidized to nitrite by the biofilm bacteria of the mains. Hence, better control of monochloramine generation to avoid excess ammonia entering the system will lower the concentration of nitrite in the distribution-supply system.

REFERENCES

Alwan, S. M., Jawad, I. M., Abdul-Rahman, S. K. and Al-Kafaji, S. H. (1987). Estimation of nitrite, nitrate and N-nitrosamines in drinking water and soft drinks. *J. Biol. Sci. Res.* **18**, 133–142.

Belser, L. and Mays, E. L. (1982). Use of nitrifier activity measurements to estimate the efficiency of viable nitrifier counts in soils and sediments. *Appl. Environ. Microbiol.* **43**, 945.

Challis, B. C. and Challis, J. A. (1982). N-nitrosamines and N-nitrosimines. In: *The Chemistry of Amino, Nitroso and Nitro Compounds and their Derivatives*. Ed. S. Patai, Wiley, New York, pp. 1151–1223.

Costerton, J. W., Cheng, K.-J., Geesey, G. G., Ladd, T. I., Nickel, J. C., Dasgupta, M. and Marrie, T. J. (1987). Bacterial biofilms in nature and disease. *Ann. Rev. Microbiol.* **41**, 435–464.

Douglass, M. L., Kabacoff, B. L., Anderson, G. A. and Cheng, M. C. (1978). The chemistry of nitrosamine formation, inhibition and destruction. *J. Soc. Cosmet. Chem.* **29**, 581–606.

Eisenbrand, G., Spiegelhalder, B. and Preussman, R. (1980). Nitrate and nitrite in saliva. *Oncology* **37**, 227–231.

Ellwood, D. C., Keevil, C. W., Marsh, P. D., Brown, C. M. and Wardell, J. N. (1982). Surface associated growth. *Phil. Trans. Roy. Soc. London B.* **297**, 517–532.

Fan, A. M., Willhite, C. C. and Book, S. A. (1987). Evaluation of the nitrate drinking water standard with reference to infant methaemoglobinaemia and potential reproductive toxicity. *Reg. Toxicol. and Pharmacol.* **7**, 135–148.

Feben, D. (1935). Nitritying bacteria in water supplies. *JAWWA* **27**, 439–447.

Finch, C. A. (1948). Methaemoglobinaemia and sulfhaemoglobinaemia. *N. Engl. J. Med.* **239**, 470–478.

Globus, M. and Samual, D. (1978). Effect of maternally administered sodium nitrite on hepatic erythropoiesis in fetal CD-1 mice. *Teratology* **18**, 367–378.

Gwatkin, R. and Plummer, P. J. G. (1946). Toxicity of certain salts of sodium and potassium for swine. *Can. J. Comp. Med.* **10**, 183–190.

Hartman, P. E. (1982). Nitrates and nitrites: ingestion, pharmacodynamics and toxicology. In: *Chemical Mutagens Vol 7*. Eds F. J. de Serres and A. Hollaender, Plenum, p. 211.

Henderson, W. R. and Raskin, N. H. (1972). 'Hot-dog' headache: individual susceptibility to nitrite. *Lancet* **2**, 1162–1163.

Höyem, T. (1974). Nitrate and nitrite contents in Norwegian food. *Proc. IV Int. Congr. Food Sci. Technol.* **3**, 466–470.

Jägerstad, M. and Nilsson, R. (1976). Intake of nitrate and nitrite of some Swedish consumers as measured by the duplicate portion technique. In: *Proceedings of the 2nd International Symposium on Nitrite Meat Products*, Zeist, pp. 283–286, Pudoc, Wageningen, The Netherlands.

Jägerstad, M. and Nordén, Å. (1976). Dietary intake of nitrate and nitrite using the duplicate sampling portion technique. *Ambio.* **6**, 276–277.

Joint Iran–International Agency for Research on Cancer Study Group (1977). Oesophageal cancer studies in the Caspian littoral of Iran. Results of population studies — a prodrome. *J. Natl. Cancer Inst.* **58**, 1127–1138.

Keevil, C. W., Glenister, D. A., Salamon, K. E., Dennis, P. J. and West, A. A. (1988). A continuous culture biofilm model for the study of medical and industrial corrosion. In: *Biofilms*. Ed. L. H. G. Morten and A. H. L. Chamberlain, Biodeterioration Society, Ken, pp. 48–62.

Keevil, C. W., West, A. A., Walker, J. T., Lee, J. V., Dennis, P. L. J. and Colbourne, J. S. (1989). Biofilms: detection, implications and solutions. In: *Proceedings of the IAWPRC Vol II*. Eds D. Wheeler, M. L. Richardson and J. Bridges, Pergamon Press, Oxford. pp. 367–374.

Keevil, C. W., Mackerness, C. W. and Colbourne, J. S. (1990). Biocide treatment of biofilms. *Int. Biodet.* **26**, 169–179.

Knight, T. M., Forman, D., Al Dabbagh, S. A. and Doll, R. (1987). Estimation of dietary intake of nitrate and nitrite in Great Britain. *Fd. Chem. Toxic.* **25**, 277–285.

Larson, T. E. (1939). Bacterial corrosion and red water. *JAWWA* **31**, 1186–1196.

Leach, S. A. (1988). Mechanisms of endogenous N-nitrosation. In: *Nitrosamines: Toxicology and Microbiology*. Ed. M. J. Hill, Ellis Horwood, Chichester, pp. 69–87.

Leach, S., Mackerness, C. W., McPherson, K., Packer, P., Hill, M. and Thompson, M. H. (1990). Some factors affecting N-nitroso compound formation from ingested nitrate in the stomach: achlorhydria, bacterial N-nitrosation and its modulation. In *Relevance to human cancer of N-nitroso compounds* (ed I. O'Neill) IARC, Lyon, p. 571–577.

Le Chevalier, M. W., Cawthorn, C. D. and Lee, R. G. (1988). Inactivation of biofilm bacteria. *Appl. and Environment Microbiol.* **54**, 2492–2499.

Lijinsky, W. (1984). Induction of tumours in rats by feeding nitrosatable amines together with sodium nitrite. *Fd. Chem. Toxic.* **22**, 715–720.

Lijinsky, W., Kovatch, R. and Riggs, C. W. (1983). Altered incidences of hepatic and hemopoietic neoplasms in F344 rats fed sodium nitrite. *Carcinogenesis* **4**, 1189–1191.

Mackawa, T., Ogiu, T., Onodera, H., Furuta, K., Matsuoka, C., Ohno, Y. and Odashima, S. (1982). Carcinogenicity studies of sodium nitrite and sodium nitrate in F-344 rats. *Fd. Chem. Toxic.* **20**, 25–33.

Mackerness, C. W., Leach, S. A., Thompson, M. H. and Hill, M. J. (1989). The inhibition of bacterially mediated N-nitrosation by vitamin C: relevance to the inhibition of endogenous N-nitrosation in the achlorhydric stomach. *Carcinogenesis* **10**, 397–399.

Mackerness, C. W., Colbourne, J. S. and Keevil, C. W. (1990). An *in vitro* chemostat model of drinking water supply systems to assess the metabolism activity of biofilm bacteria (in press).

Ministry of Agriculture, Fisheries and Food. The twentieth report of the steering group on food surveillance. *The Working Party on Nitrate and Related Compounds in Food (1987). Nitrate, Nitrite and N-nitroso Compounds in Food.* Food Surveillance Paper No. 20. HMSO, London.

Mirvish, S. S. (1983). The aetiology of gastric cancer — intragastric nitrosamide formation and other theories. *J. Nat. Cancer Inst.* **71**, 629–647.

Mirvish, S. S., Wallcave, L., Eagen, M. and Shubik, P. (1972). Ascorbate nitrite reaction: possible means of blocking the formation of carcinogenic N-nitroso compounds. *Science* **177**, 65–68.

Mirvish, S. S., Bulay, O., Runge, R. G. and Patil, K. (1980). Study of the carcinogenicity of large doses of dimethylnitramine N-nitroso-1-proline and sodium nitrite administered in drinking water to rats. *J. Natl. Cancer Inst.* **64**, 1435–1442.

Naida, S. R. and Venkatrao, P. (1945). The toxicology of nitrites. *Calcutta Med. J.* **42**, 79–87.

National Academy of Sciences (NAS) (1981). The health effects of nitrate, nitrite and N-nitroso compounds. National Academy of Sciences, Washington DC.

Oshima, H. and Bartsch, H. (1981). Quantitative estimation of endogenous nitrosation in humans by monitoring N-nitrosoproline excreted in the urine. *Cancer Research* **41**, 3658–3662.

Ralt, D. and Tannenbaum, S. R. (1981). The role of bacteria in nitrosamine formation. In: *ACS Symposium series 174.* Eds R. A. Scanlan, and S. R. Tannenbaum, American Chemical Society, Washington, pp. 159–164.

Reed, P. I., Smith, P. L. R., Haines, K., House, F. R. and Walters, C. L. (1981). Gastric juice N-nitrosamines in health and gastroduodenal disease. *Lancet* **ii**, 550–552.

Rittman, B. E. and Snoeyink, V. L. (1984). Achieving biologically stable drinking water. *JAWWA* **76**, 106–114.

Rodkey, F. L. (1976). A mechanism for the conversion of oxyhaemoglobin to methaemoglobin by nitrite. *Clin. Chem.* **22**, 1986–1990.

Roth, A. C., Herkert, G. E, Bercz, J. P and Smith, M. K. (1987). Evaluation of the developmental toxicity of sodium nitrite in Long–Evans rats. *Fundamental and Appl. Toxicol.* **9**, 668–677.

Ruddell, W. S. J., Bone, E. S., Hill, M. J., Blendis, L. M. and Walters, C. L. (1976). Gastric juice nitrite: a risk factor for cancer in the hypochlorhydric stomach. *Lancet* **ii**, 1037.

Ruddell, W. S. J., Blendis, L. M. and Walters, C. L. (1977). Nitrite and thiocyanate found in the fasting and secreting stomachs and in saliva. *Gut* **18**, 73–77.

Sander, J. (1971). Untersuchengen über die Entstehung cancerogener Nitrosonerbindung im Magen Von Versuchstieren and ihre Bedeutung für die Menschen. *Arzneim-Forsch* **21**, 1572–1580, 1707–1713, 2034–2039.

Schweinsberg, F. and Bürkle, V. (1985). Nitrite: a co-carcinogen? *J. Cancer Res. Clin. Oncol.* **109**, 200–202.

Selenka, F. and Brand, E. (1975). Nitrate and nitrite in prepared meals in relation to the nitrate concentration in drinking water. *Zbl. Bakt. Hyg. I. Abt. Orig. B.* **161**, 266–279.

Selenka, F. and Brand-Grimm (1978). Nitrat und Nitrit in der Ernährung des Menschen: Kalkulation der mittleren Tagesanfriahme und Abschätzung der Schwankingsbreite. *Zbl. Bakt. Parasitenkd. Infektionskr. Hyg. Abt. I. Orig. Reihe. B.* **162**, 449–466.

Shimada, T. (1989). Lack of teratogenic and mutagenic effects of nitrite on mouse fetuses. *Arch. Environmental Health* **44**, 59–63.

Shuker, D. E. G. (1988). The chemistry of N-nitrosation. In: *Nitrosamines: Toxicology and Microbiology*. Ed. M. J. Hill, Ellis Horwood, Chichester, pp. 48–68.

Shuval, H. I. and Gruener, N. (1972). Epidemiological and toxicological aspects of nitrates and nitrites in the environment. *Am. J. Public Health* **62**, 1045–1052.

Sollman, T. (1957). *A Manual of Pharmacology*, 8th edn. W. B. Saunders, Philadelphia.

Stephany, R. W. and Schuller, P. L. (1978). The intake of nitrate, nitrite, volatile –N-nitrosamines and the occurrence of volatile N-nitrosamines and the occurrence of volatile N-nitrosamines in human urine and veal calves. In: *Environmental Aspects of N-nitroso compounds*. Eds E. A. Walker, M. Castegnaro, L. Griciute, R. E. Lyle and W. Davis, IARC Scientific Publication No. 19, pp. 443–460. International Agency for Research on Cancer, Lyon, France.

Stouthamer, A. H. (1976). Biochemistry and genetics of nitrate reductase bacteria. *Adv. in Microbial Physiology*. Eds A. H. Rose and D. W. Tempest, pp. 315–375.

Stuehr, D. J. and Marletta, M. A. (1987). Synthesis of nitrate and nitrite in murine macrophage cell lines. *Cancer Research* **47**, 5590–5594.

Suzuki, I. (1974). Mechanisms of inorganic oxidation and energy coupling. *Ann. Rev. Microbiol.* **28**, 85–102.

van Loosdrecht, M. C. M., Lykleman, J., Norde, W. and Zehnder, A. J. B. (1990). Influence of interfaces on microbial activity. *Microbiological Reviews* **54**, 75–87.

Vorhees, C. V., Butcher, R. E., Brunner, R. L. and Wootten, V. (1984). Developmental toxicity and psychotoxicity of sodium nitrite in rats. *Food Chem. Toxic.* **22**, 1–6.

Walters, C. L. (1980). The exposure of humans to nitrite. *Oncology* **37**, 289–296.

Walters, C. L., Smith, P. L. R, Reed, P. I., Haines, K. and House, F. R. (1982). N-nitroso compounds in gastric juice and their relationship to gastroduodenal disease. In: *N-nitroso Compounds, Occurrence and Biological Effects*. Eds. H.

Bartsch, I. K. O'Neill, M. Castegnaro and M. Okoka, IARC Scientific Publication No. 41. International Agency for Research on Cancer, Lyon, pp. 345–355.

White, J. W. (1975 and 1976). Relative significance of dietary sources of nitrate and nitrite. *J. Agric. Food Chem.* **23**, 886–891 (1975) and **24**, 202 (1976).

Winogradsky, A. (1949). Contribution on l'étude de la microflore nitrificatrice des eaux usées: resistance des germs aux conditions défavorables. *Ann. Inst. Pasteur.* **76**, 35–42.

Woolfe, R. L., Means, E. G., Davis, M. K. and Barrett, S. E. (1988). Biological nitrification in covered reservoirs containing chloraminated water. *JAWWA* **80**, 109–114.

Yang, C. S. (1980). Research on oesophageal cancer in China: a review. *Cancer Res.* **40**, 2633–2644.

5

Nitrate and nitrate in foods

C. L. Walters

5.1 THE NITROGEN CYCLE

The primary source of nitrogen is that of the element in a gaseous form in the atmosphere. Its incorporation into terrestrial nitrogenous compounds such as ammonia takes place along a number of different pathways, including micro-organisms, plants and man himself through his industrial and agricultural activities; electrical discharges such as lightning can also contribute to the combination of nitrogen and oxygen. It has been estimated (WHO 1977) that the 'fixation' of atmospheric nitrogen amounts to 150 million tonnes of the element each year. Of that, industrially fixed nitrogen amounts to 30 million tonnes in 1970 and is expected to exceed 100 million tonnes by the year 2000 (Ellen and Schuller 1983).

Bacteria of the *Rhizobium* species present in the root nodules of legumes such as beans, clover, alfalfa and soya contribute strongly to the so-called nitrogen cycle in reducing elemental nitrogen to ammonia. Its conversion to nitrate (nitrification) proceeds via the intermediate formation of nitrite through a two-stage process involving different microorganisms.

Many plants can assimilate and concentrate nitrate, only a part of which can be utilized. The nitrogen involved can be incorporated into proteins and probably other nitrogenous constituents or can act as a terminal electron acceptor to the respiratory chain of the chloroplasts. The degradation of nitrate (denitrification) to nitrogen and/or oxides of nitrogen takes place in the soil and also lakes and other water courses. Nitrates from natural fixation and that applied as a fertilizer can be used for the synthesis of biological components and particularly proteins. Waste material from plants and animals returns combined nitrogen to the soils in which part is recycled and part returned to the atmosphere to complete the nitrogen cycle (Fig. 5.1).

Nitrate is necessary for good plant growth. According to Christy *et al.* (1973), probably more than 90% of the nitrogen absorbed by plants is in the form of nitrate. The nitrate content of plants is governed by inherited genetic factors controlling their

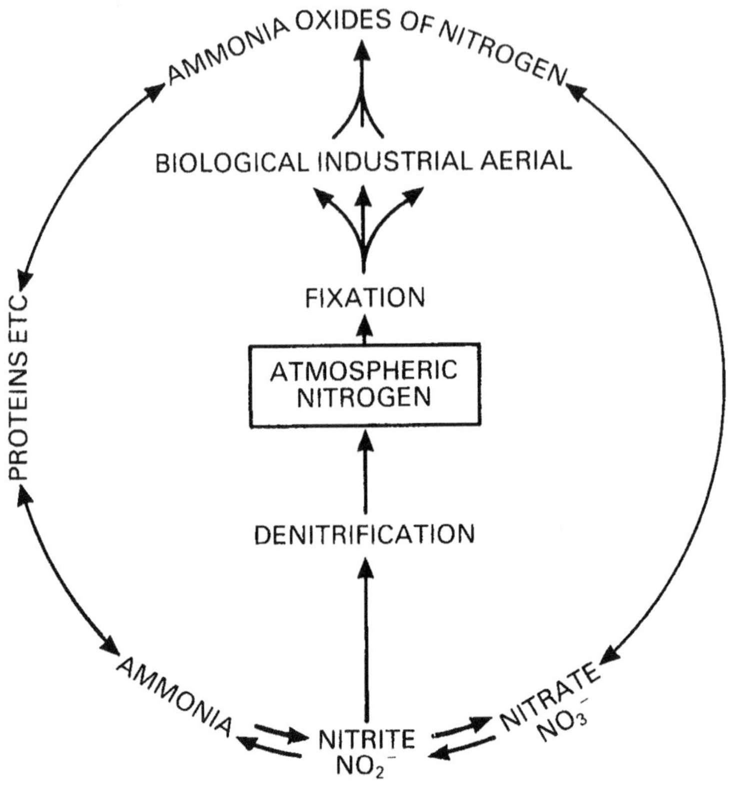

Fig. 5.1 — Elements of the nitrogen cycle.

metabolic patterns and the quantity of nitrate available in the soil. Leafy green vegetables and some root crops assimilate nitrate naturally, and their contents are discussed in a later section of this chapter; wide variations in nitrate concentrations occur between species. Fertilizer applied to excess beyond the capabilities of plants to use them may result in the accumulation of nitrate. Throughout the world, nitrogen fixation may exceed denitrification by about 10% according to Delwiche (1970).

5.2 THE STABILITIES OF NITRATE AND NITRITE IN FOODS

In food terms, nitrate is very stable chemically throughout the relevant range of pH. It can be reduced to nitrite, however, in contact with metals such as occurs during the cooking of food in aluminium utensils (WHO 1977). Many plants and micro-organisms contain enzymes' systems capable of utilizing nitrate with the production of nitrite, oxides of nitrogen, nitrogen and even ammonia. Of recent times, evidence has surfaced of the presence of nitrate reductase enzyme systems in some mammalian tissues such as parts of the gastrointestinal tract (Ward *et al.* 1986). Whether or not fresh meat in the form of skeletal muscle is capable of reducing nitrate, however, has not yet been resolved, but it seems unlikely since its constituents, myoglobin and haemoglobin, are not converted into their nitrosyl forms during storage, as would have occurred if nitrite had been formed. Conversely, the storage of spinach in the presence of bacteria promotes the ready reduction of part, at least, of its nitrate content to nitrite.

On the other hand, nitrite is unstable, particularly at acidic pH values at which it can disproportionate to yield nitrate and nitrogen oxide and/or react with many components of foods including amines, phenols and thiols. The products can include elemental nitrogen from primary amines, N-nitroso compounds from secondary and tertiary amines, C-nitroso and -nitrophenols from phenols, and S-nitrosothiols from thiols.

Many fruits and vegetables contain ascorbic acid, and this vitamin and its iso form, erythorbic acid, have been used as an aid in the manufacture of cured meats such as bacon. Ascorbic acid reacts readily with nitrite, a major product in the absence of oxygen being nitrogen oxide (Walters and Taylor 1964). Where oxygen is present, nitrogen dioxide, a powerful nitrosating agent of secondary amines and other receptors, is produced, although the reaction is a trimolecular one involving two moles of nitrogen oxide and one mole of oxygen; at low levels of nitrogen oxide, therefore, its combination with oxygen proceeds surprisingly slowly, and very low concentrations of nitrogen oxide are virtually stable in air as a consequence. Amongst other scavengers of nitrite present in foods are the antioxidants, such as α-tocopherol. Great care must be exercised in determining nitrite in the presence of such scavengers with which it can react at the acidic pH values required for its measurement as a diazonium salt.

The smoking of fish or meat and direct-fired drying of such a commodity as milk can be accompanied by the formation of oxides of nitrogen leading to the deposition of nitrate in the foods. For instance, the nitrate nitrogen contents of direct-fired dried milk products were raised by 1–3 mg/kg compared with those prepared using indirectly heated sprayers, according to Manning *et al.* (1968); somewhat surprisingly, nitrite nitrogen levels were unaffected, possibly by virtue of its ready reaction with food components.

5.3 CONVERSION OF NITRATE TO NITRITE IN FOODS

Spinach is one of the vegetables best able to accumulate nitrate from the soil, with concentrations often in thousands of milligrams per kg. As such, its conversion to nitrite during storage has been the subject of particular scrutiny in view of the ability of this product to oxidize oxyhaemoglobin to the met form, which is incapable of transporting oxygen and to act as a precursor to carcinogenic N-nitroso compounds.

According to Phillips (1968a), the initial nitrite concentrations of fresh, frozen, canned and baby-food spinach samples were generally less than one milligram nitrite nitrogen per kilogram fresh weight. Storage at room temperature, however, resulted in rapid falls in the nitrate contents and concomitant increases in the nitrite concentrations. When refrigerated, the nitrite nitrogen contents of fresh spinach increased very gradually during a storage of 28 days (Phillips 1968a). No significant increases in nitrite contents occurred during the storage of frozen, canned or baby-food spinach, but increased concentrations were observed in frozen spinach left to thaw at room temperature for 39 h. A small rise in nitrite levels was also apparent when partially consumed jars of commercial baby foods containing nitrate were stored for seven days at room temperature rather than under refrigeration.

Nitrate has been employed as a food additive to cured meats over many years. Before cooking, residual mammalian enzymes remain active, but no evidence has been provided that a nitrate reductase system(s) is included, as has been found in rat stomach and small intestine (Ward *et al.* 1986). Nevertheless, cured meats are often contaminated with bacteria, some of which at least are able to reduce nitrate to nitrite. This conversion proceeds more rapidly in unpackaged bacon than in the vacuum-packed variety (Cavett 1962), a difference which has been ascribed somewhat surprisingly to the low reducing activity of anaerobic bacteria.

Cooking in aluminium utensils may bring about the reduction of nitrate to nitrite (WHO 1977), an effect which could be of concern in countries in which this metal is used in vessels for the boiling of milk or water. The subsequent decay of nitrite formed depends to a great extent upon the pH of the product and the temperature of storage.

5.4 NITRATE AND NITRITE CONTENTS OF FOODS

5.4.1 Nitrate and nitrite in vegetables

Nitrate concentrations in vegetables vary enormously, ranging from around 1 to 10 000 mg/kg fresh weight (MAFF 1987). Even amongst different samples of the same variety of vegetables the range of concentrations can be very large. For instance, the nitrate concentrations in beetroot from various retail outlets varied from 630 to 6800 mg/kg within the same investigation (Pickston *et al.* 1980). Beetroot, celery, lettuces, spinach and radishes are amongst the greatest sources of nitrate, often exceeding 1000 mg/kg, whereas broad beans, peas, cauliflowers, potatoes, onions and sweetcorn generally contain less than 200 mg/kg (MAFF 1987). A classification of vegetables in relation to their nitrate contents was made by Corré and Breimer (1979). Thus, nitrate contents of Group 1 are mainly less than 200 mg NO_3^- ion/kg; this group includes asparagus, mushrooms, peas, peppers, white and sweet potatoes and tomatoes. Vegetables within Group 2 have nitrate concentrations mostly of less than 500 mg NO_3^-/kg, and encompass broccoli, cauliflower, cucumber, onions and turnips as well as melons. Nitrate levels of up to 1000 mg NO_3^-/kg cover the vegetables of Group 3 within which are to be found green beans, cabbages, carrots and kale. Endives, leeks, parsley and rhubarb have nitrate contents which are mostly of less than 2500 mg NO_3^-/kg, and constitute Group 4.

Vegetables with the highest nitrate contents of generally higher than 2500 mg NO_3^-/kg include beets, celery, lettuce, spinach and radishes which are classed as belonging to Group 5. Table 5.1 records the ranges of nitrate concentrations presented by the Committee on Nitrite and Alternative Curing Agents in Food (US National Academy of Sciences 1981).

Table 5.1 — Ranges of nitrate contents of fresh, canned or frozen vegetables

Vegetables	Ranges of nitrate contents mg NO_3^-/kg	Class†
Asparagus	3–700	1
Beet	100–4500	5
Broccoli	140–2300	2
Cabbage	0–2700	3
Carrot	0–2800	3
Cauliflower	53–4500	2
Celery	50–5300	5
Cucumber	17–570	2
Kale	30–5500	3
Lettuce	90–13000	5
Parsley	0–4100	4
Peas	20–100	1
Potato (white)	57–1000	1
Radish	60–9000	5
Spinach	2–6700	5
Tomato	0–170	1

† See text for classification.
From *The Health Effects of Nitrate, Nitrite and N-nitroso Compounds* (US National Academy of Sciences, 1981)

Although the analytical procedures employed for the determination of nitrate have varied from study to study, the wide ranges observed for many vegetables are not considered to accrue from such diversity but from actual variations in their contents within the same types of vegetable. Changes in the nitrate contents of vegetables over the period from 1907 to 1970 were studied by Maynard and Barker (1972), who concluded that no tendency for substantial change was reflected in their data. Nevertheless, a comparison between the average concentrations of nitrate of fresh market vegetables in the 1960s and 1970s based on the data of Corré and Breimer (1979) suggests that increases had occurred in some vegetables, including carrots, lettuce and spinach. Mean values for carrots had risen from 130 mg NO_3^-/kg (5 samples) to 330 mg/kg (19 samples), those for lettuce from 1100 mg/kg (5 samples) to 2800 mg/kg (33 samples) and for spinach from 1100 mg/kg (11 samples) to 1900 mg/kg (34 samples) over the decade.

The use of inorganic fertilizers has been increasing over recent decades, and Barker *et al.* (1971) came to the conclusion that the nitrate contents of vegetables tended to increase with the amount of nitrogen applied, whether or not the crop yield was influenced. Mills *et al.* (1976) fertilized a high-nitrate spinach with 50% ammonium nitrogen + 50% nitrate nitrogen, instead of 100% nitrate nitrogen and thereby reduced its nitrate content from 90 g NO_3^-/kg dry weight to 60 g NO_3^-/kg dry weight, i.e. a 34% reduction. When nitropyrin was used with a 1:1 mixture of ammonium nitrogen and nitrate nitrogen, a 34% reduction was again observed in the spinach nitrate content.

Amongst other factors influencing the nitrate content of vegetables are the choice of cultivar, i.e. the strain of Rybrid. As an example, the nitrate concentration of the smooth-leaved variety of spinach (cv. Tuftegard) was 2 g NO_3^-/kg dry weight, according to Cantliffe (1972), compared with a level of 7.4 g NO_3^-/kg on the same basis for the savoyed-leaved variety (cv. Bloomsdale).

Although nitrate uptake by plant roots is not markedly affected by photosynthesis, its reduction and assimilation are closely coupled with the effect of light on such processes proceeding predominantly in the leaves (US National Academy of Sciences 1981). Nitrate is metabolized by plant nitrate reductase enzymes, and thus accumulation can occur where such enzyme activity is inhibited or prevented. Conversely, leafy plants harvested on a sunny afternoon favouring photosynthesis frequently contain less nitrate than those collected in the morning or during inclement weather.

The geographical region and the particular season of harvest involved can be instrumental in influencing nitrate levels, which tend to increase as daytime temperatures fall below an optimal but unspecified temperature (US National Academy of Sciences 1981). As an example, greenhouse carrots grown in winter had higher nitrate levels than those harvested in summer (Habben 1973), perhaps as a result of lower light intensities in winter. This proposition is supported by the higher nitrate contents in beets illuminated for 8 h periods than those irradiated with 20 h periods (Cantliffe 1972). Similarly, plants denied sunlight by being grown in shade or at high latitudes and under drought conditions conserve higher levels of nitrate than those produced under optimal conditions of humidity and light. The higher levels of nitrate accumulated by greenhouse plants in comparison with those grown outdoors may well result from the greater use of nitrogen fertilizers indoors.

Other factors which can result in nitrate accumulation include some plant diseases, insect damage or contact with herbicides, such as those used in weed control (US National Academy of Sciences 1981). Nitrate reductase enzymes are dependent on the availability of molybdenum, and plants grown in soils deficient in this essential element or potassium can have higher nitrate contents than those nutritionally replete. Growth in organically rich peat soils can also lead to elevated nitrate contents whilst increasing the irrigation of growing plants may reduce the concentrations at harvest (Carter and Bosma 1974). According to Vulsteke and Biston (1978), nitrate levels in snap beans, carrots and celery decreased with increasing maturity at the time of harvest.

Factors influencing the accumulation of nitrate in plants can interact synergistically. For instance, the nitrate content (approximately 0.4 mg NO_3^-/kg fresh weight) of tomatoes from plants grown at a daylight temperature of 35°C was similar whether or not nitrogen fertilizer was added. Nevertheless, that from plants grown at 20°C was restricted to a content of approximately 0.7 mg NO_3^-/kg when excess nitrogen was denied, but reached about 73 mg/kg when excess nitrogen was applied. As well as increasing with the amount of nitrogen supplied, nitrate levels generally rise when the fertiliser is applied shortly before harvest.

The German Democratic Republic, Switzerland and the Netherlands have introduced measures to limit the nitrate content of leafy vegetables in particular. Currently the permitted levels are high but it is proposed to lower them as it becomes commercially practicable with time.

Great care must be exercised in the determination of nitrite in a biological matrix because of its instability, particularly at acidic pH values. Apart from its disproportionateness to nitrate and nitrogen oxide, nitrite reacts readily with phenols and other constituents of many foods. The determination of nitrite through the formation of a dye requires the acidification of the reaction medium, with the possibility of losses through interaction with co-extracted compounds. Ascorbic acid, for instance, reacts with nitrite over a wide range of pH values (Walters and Taylor 1964), and thus it should be removed initially by, for instance, its inactivation by titration with iodine before the formation of an azo dye. Some of the labile compounds which can be formed from nitrite in a biological matrix would not have contributed directly to the values observed for the nitrite contents of foods unless means for their decomposition had been included in the assays. For instance, the S-nitroso derivatives of thiols can be decomposed to form nitrite using iron salts. Nevertheless, such compounds can be important in relation to the formation of carcinogenic N-nitroso compounds, particularly at pH values around neutrality and above (Massey *et al.* 1980).

The nitrite content of most fresh market vegetables is low and usually of the order of 1–2 mg NO_2^-/kg (Corré and Breimer 1979). Siciliano *et al.* (1975), for instance, concluded that the nitrite contents of commercial fresh, frozen or canned vegetables, as purchased by the consumer, are generally less than 1 mg NO_2^-/kg. Surprisingly, higher values (4.3–76 mg NO_2^-/kg) were reported by Richardson back in 1907, for which no reason can be advanced.

5.4.2 Effects of storage on nitrate and nitrite in vegetables

Contradictory reports have been issued on the effects of storage on the contents of nitrate in vegetables. According to Japanese workers (Hata and Ogata 1971a), the nitrite level in carrots did not change during storage whilst potatoes kept under a low partial pressure of oxygen showed lower nitrate and higher nitrite concentrations (Hata and Ogata 1971b). Similarly, studies in New Zealand (Pickston *et al.* 1980) on a variety of vegetables gave no consistent changes in their nitrate and nitrite contents.

Nevertheless, other workers (Corré and Breimer 1979, Lee 1970, Lin and Yen 1980, Phillips 1968a,b 1971) contend that the nitrite contents of vegetables increase on storage, especially at higher temperatures as a result of the reduction of nitrate by plant or bacterial enzymes. This is especially true of processed vegetables, possibly as a result of the release of nitrate from vacuoles in which it is normally isolated

(Martinoia *et al.* 1981). As a consequence, the nitrite contents of processed or even frozen vegetables can be two to threefold greater than those of their unprocessed counterparts (Corré and Breimer 1979, Siciliano *et al.* 1975); processed infant foods can also contain elevated levels of nitrite (Kamm *et al.* 1965). As an extreme example, the nitrite content of vegetables deliberately preserved by fermentation, as in delicacies consumed in parts of China and Japan, can reach 400 mg NO_2^-/kg (Matsui 1944, Yanagihara *et al.* 1963).

5.4.3 Ascorbate in vegetables

Reference has already been made to the action of ascorbate, which occurs in many vegetables, as a nitrite scavenger. It is also important in that it can inhibit markedly the formation of carcinogenic N-nitroso compounds from constituent amines, amides, guanidines, etc. Schuphan (1974) and Li *et al.* (1980) observed an inverse relationship between nitrate and ascorbate concentration in vegetables, thereby supporting the findings of Kilgore *et al.* (1964). They reported that turnip greens from unshaded plants treated with normal levels of nitrogenous fertilizer contained 1593 mg NO_3^-/kg and 1351 mg ascorbate/kg whereas shaded plants exposed to excess fertilizer and sodium nitrate contained 4707 mg NO_3^-/kg and only 833 mg ascorbate/kg. Thus the ascorbate:nitrate ratios were 0.30 and 0.06 respectively on a molecular basis.

According to a report by the Committee on Nitrite and Alternative Curing Agents in Foods (Assembly of Life Sciences 1981), the molar ratios of ascorbate to that of nitrate for vegetables and fruits ranged from 0.02 for beets, celery and rhubarb, 0.03 for lettuce, 0.05 for radish and 0.10 for spinach to 3.0 for okra, 3.1 for brussels sprouts, 3.5 for artichokes and 3.7 for sweet peppers. The levels of ascorbate quoted were those for fresh, uncooked produce as purchased. For a few vegetables, therefore, sufficient ascorbate is available to prevent any potentially adverse reactions by nitrite even if all of their nitrate contents had been reduced to this nitrosating agent by plant or bacterial enzymes or other means. It is unlikely, however, that the ascorbate content of nitrate-rich vegetables, such as beets, celery, lettuce, radish and spinach, would be sufficient to decompose nitrite formed to any great extent from the conversion of available nitrate. Thus, any spinach products for infant nutrition in particular should be stored under hygienic conditions since little protection from methaemoglobinaemia can be afforded by its ascorbate content *per se*.

5.4.4 Nitrate and nitrite in fruit

Nitrate contents of fruit are generally low and of the order of 10 mg NO_3^-/kg whilst nitrite is virtually absent according to reviews by White (1975, 1976) of three previous reports. However, strawberries sometimes contain in excess of 100 mg/kg of the nitrate ion according to Herrmann (1972), and grapes can reach a level of 17 mg/kg from time to time. Nitrate levels in pears (34 mg NO_3^-/kg), cherries (24 mg NO_3^-/kg) and apples 24 mg NO_3^-/kg) found by Huguet *et al.* (1976) were not influenced by the addition of fertilizer, but Kenny and Walshe (1975) failed to detect nitrate in Golden Delicious apples. Nitrite levels reported (Harada *et al.* 1972) in apples, oranges and other fruits were less than 1 mg NO_2^-/kg. Thus, it would appear that the contribution of fruit to the human intake of nitrate is small whilst that to the nitrite burden is negligible.

5.4.5 Nitrate and nitrite contents in meat

In common with other biological tissues, skeletal muscle is considered to contain nitrate distributed from that ingested by the animal in its forage or drink. Much of the intake of nitrate is excreted in the urine and other secretions, but the remainder is probably transported in the blood to the tissues. Where bacteria are available, as in the rumen or resulting from a urinary tract infection, nitrate will be degraded to a range of products.

Most levels of nitrate determined in fresh meat products have been low. As an example, a value of 0.9 mg NO_3^-/kg was reported by Wright and Davison (1964) for the meat of dairy cows studied as controls in a feeding experiment. However, Usher and Telling (1975) concluded that nitrate concentrations in blank determinations in a series of studies ranged from 0 to 49 mg NO_3^-/kg. Kacmar and Bartik (1965) reported even higher nitrate concentrations of 170 mg NO_3^-/kg in fresh pork muscle, but these would appear to be exceptional.

Little information is available on the nitrite contents of fresh meat products. If formed from available nitrate, it would be reduced to nitrogen oxide either free or in combination with endogenous myoglobin by enzymes such as those of the muscle respiratory chain (Walters *et al.* 1967). Nevertheless, dry-cured hams processed with only salt and sugar were claimed to contain nitrite to approximately 5 mg NO_2^-/kg (Kemp *et al.* 1975), a value which surprisingly was not altered markedly on storage.

The preservation of meat using salt has been a commonly occurring practice throughout the world for centuries and was designed to provide a continuing supply at times of scarcity. Most sources of salt contained nitrate as an impurity. However, it was not until this century that it was realized that it was nitrite derived from nitrate by microbiological reduction that was responsible for much of the curing action of the brine with which the meat was processed.

Subsequently, nitrite itself was employed as the active ingredient and its concentration was controlled initially to promote an adequate level of conversion of endogenous myoglobin to its red nitrosyl form and to avoid the possibility of methaemoglobinaemia in the consumers. Initially the use of nitrate in the processing of Wiltshire bacon was unlimited, since it was considered to be without toxic effect and to provide a source of available nitrite as its level declined during storage. Of recent times, no less than 7% of all food consumed in the USA had been prepared using nitrite (Anon 1980).

Currently the main *raison d'être* for the continued use of nitrite and nitrate rests on the ability of their breakdown products to inhibit the outgrowth of spores of *Clostridium botulinum* naturally occurring in foods and thereby to prevent the production of its toxin. This permission persists in spite of the fact that nitrite is recognized to act as a precursor to potentially carcinogenic N-nitroso compounds both in foods and in the stomach of the consumer.

Specific legislative provision for the use of nitrite in food has existed in the UK in one form or another since 1940 whilst nitrate was brought within the scope of legislation in 1962 (MAFF 1987). Previously, the level of nitrate permitted in Wiltshire bacon in the UK was unlimited and this could lead on occasions to residual nitrite concentrations of several hundreds or even thousands of parts per million where the curing brine was contaminated with large numbers of nitrate-reducing organisms. Under such circumstances, curing tanks could be shrouded in brown clouds of nitrogen dioxide, according to workers in the bacon industry.

Maximum permitted levels for nitrite quoted in the 1962 regulations were 200 mg/kg in cooked pickled meat and 500 mg/kg in uncooked pickled products. Amendments to the regulations in 1971 (Anon) permitted levels of sodium nitrate and sodium nitrite of up to 500 mg/kg and 200 mg/kg respectively. Current provisions arising from amendments in 1982 of the 1979 regulations (Anon) vary the permitted levels of nitrate and nitrite according to the need for microbiological protection. Thus in a sterile pack, the combined concentrations permitted are 150 mg/kg, of which not more than 50 mg/kg may consist of sodium nitrite. At the other end of the scale, uncooked bacon and ham may contain 500 mg/kg of the sodium salts of nitrate and nitrite of which the latter is restricted to 200 mg/kg. In a survey of cured-meat products in 1975 (MAFF 1987), the mean levels of nitrate ranged from 50 mg/kg for corned beef and tongue to 200 mg/kg for canned ham. The corresponding range for nitrite was from 10 mg/kg for chopped ham and pork to 35 mg/kg for bacon and ham. Overall, the maximum observed levels of nitrate in cured meats had declined from 6000 mg/kg in 1959 to 1000 mg/kg in 1972 and to around 450 mg/kg in 1975 (MAFF 1987). A similar reduction in nitrite concentrations was reported from levels frequently above 100 mg/kg before 1970 to below 50 mg/kg in 1973–4.

Similar changes in permitted levels of nitrate and nitrite have been enforced in other countries. Previously, the permitted level of addition of nitrite in the USA was ¼ oz per 100 lb of meat, equivalent to a final concentration of 156 mg/kg. From average levels of nitrate in hams in the USA of 850 mg/kg in 1907 and 600 mg/kg in 1937, its concentration had been reduced to 400 mg/kg by 1970. The corresponding figures for nitrate in bacon were 390, 1200 and 80 mg/kg in 1907, 1937 and 1970 respectively (Ellen and Schuller 1983). The addition of nitrite to meats in the Federal Republic of Germany is restricted to that of Pökelsalz, proprietary mixtures with sodium chloride of strictly controlled composition. This approach has two advantages over the direct addition of nitrite. Firstly, larger quantities of the mixture are weighed out, reducing thereby the chance of error. Secondly, any illegal addition of nitrite would be accompanied by excessive salinity, leading to a loss of palatability of the meat product.

The importance of the level of nitrite added to meat is intimately bound up with the formation of carcinogenic N-nitroso compounds from nitrosatable precursors in the substrate. In many instances, the extents of nitrosation can be reduced, but not generally prevented, by the addition of a nitrite scavenger, such as ascorbic acid, which is sometimes used as a curing aid; α-tocopherol has also proved successful as an inhibitor of nitrosation in experimental studies along with other antioxidants such as octyl gallate (Walters *et al.* 1976). Other additives used in the production of specialized meat products, such as spices, may contain constituents which are even more nitrosatable than components of the meat tissues themselves.

Of the studies of the nitrate and nitrite levels in cured meats, bacon has featured prominently by virtue of its consistent content of volatile N-nitrosamines after frying. The levels of both additives vary widely within a commodity; according to Panalaks *et al.* (1973), for instance, nitrite concentrations in 20 bacons prepared using this additive ranged from 7 to 320 mg NO_3^-/kg with a mean of 96 mg/kg. After frying, contents of 4–98 mg NO_3^-/kg were recorded for 12 samples (Panalaks *et al.* 1974). An even greater range of nitrite concentrations from less than 0.7 to 1400 mg NO_2^-/kg (mean 150 mg/kg) was reported for 23 hams cured with nitrite (Panalaks *et al.* 1973). Of continental-type sausages, the nitrate contents of 25 samples of salami ranged from 4 to 270 mg NO_3^-/kg (mean 89 mg/kg) (Panalaks *et al.* 1973) and for a further 63 samples from 4 to 540 mg NO_3^-/kg (mean 78 mg/kg) (Panalaks *et al.* 1974). Eighteen samples of other smoked and unsmoked continental sausages contained nitrate ranging from 17 to 240 mg NO_3^-/kg with a mean of 110 mg/kg (Panalaks *et al.*, 1973). It is noteworthy that British-type sausages are not prepared using nitrate or nitrite and thus the contents of these additives will be typical of fresh meat products; in fact, the sulphur dioxide content of British sausages should destroy any nitrite formed from endogenous nitrate. Similar determinations in 16 shelf-stable canned cured meats ranged from less than 0.7 mg NO_3^-/kg to 840 mg/kg (mean 100 mg/kg) (Panalaks *et al.* 1973) and for seven further samples from less then 0.7 to 110 mg NO_3^-/kg (mean 26 mg/kg) (Panalaks *et al.* 1974).

Early studies (Kerr *et al.* 1926) of the nitrite contents of 12 uncooked bacons cured in brines containing nitrate showed a range of 37–430 mg NO_2^-/kg. More recent products prepared using nitrite itself generally provided more limited ranges of concentration. Those for 12 such samples by Sen *et al.* (1974) ranged from 8 to 63 mg NO_/kg (mean 28 mg/kg) and for 14 further samples (Sen *et al.* 1975) 7 to 68 mg NO_2^-/kg (mean 35 mg/kg). Somewhat greater ranges were reported in two other studies, namely from none detected to 88 mg NO_2^-/kg (mean 42 mg/kg) (Sen *et al.* 1977) and 3–170 mg NO_2^-/kg (mean 25 mg/kg) (Panalaks *et al.*, 1973). Greenberg (1977) reported a lower average of 17 mg/kg with a range of 4–32 mg/kg for 12 samples produced in 1974 but a mean of 35 mg/kg (range 7–68 mg/kg) for a further 14 samples in 1976. After frying, a smaller range (5–18 mg NO_2^-/kg) was reported by Panalaks *et al.* (1974) with a mean of 7 mg/kg.

The concentrations of nitrite in 23 hams prepared using this additive alone ranged from less than 0.7 to 140 mg/kg (mean 29 mg/kg) (Panalaks *et al.* 1973). This compares with a range of 24–640 mg/kg in earlier studies by Kerr *et al.* (1926) in hams prepared in brines containing nitrate. Studies on the average nitrite levels in salami-type sausages by Panalaks *et al.* (1973, 1974) yielded values of 17 mg NO_2^-/kg (range of less than 0.7 to 66 mg/kg for 25 samples) and 13 mg/kg (range of 5–97 mg/kg for 63 samples). A much lower average nitrite concentration of 9.6 mg/kg (range less than 0.7–52 mg/kg) was provided by 18 smoked and unsmoked continental sausages than that reported (190 mg NO_2^-/kg) much earlier by Kerr *et al.* (1926) for 13 similar samples prepared using nitrate. A low average value of 6 mg NO_2^-/kg was reported from two studies by Panalaks *et al.* (1973, 1974) firstly on 16 samples (range less than

0.7–17 mg/kg) and subsequently on seven samples (range 5–8 mg/kg) of shelf-stable canned cured meats.

The use of nitrate and nitrite as food additives has been virtually abolished in Norway in recent years although it is reputed that wherever possible meat manufacturers have replaced nitrite with oxides of nitrogen formed during the smoking process. Nevertheless, no less than 65% of the 297 Canadian meat products produced conventionally and analysed by Panalaks *et al.* (1973, 1974) showed nitrite concentrations of only up to about 10 mg NO_2^-/kg, with 127 products (43%) containing levels of 7 mg/kg or less, a level not dissimilar from the average for Norwegian products. A similar ban in Canada in 1975 on the addition of nitrate to all but a few products in which it was deemed necessary has resulted in a fivefold decrease in its content in cured meats between 1971 and 1978, according to a review by the Committee on Nitrite and Alternative Curing Agents in Foods (US National Academy of Sciences 1981). Thus the average residual nitrate concentration in retailed Canadian meat products declined from 130 mg NO_3^-/kg in 1971 to 64 mg/kg in 1973 and to 28 mg/kg in 1978. The corresponding values for residual nitrite levels were 19, 11 and 15 mg NO_2^-/kg for surveys in 1971, 1973 and 1978 respectively, thereby suggesting that its formation from nitrate *in situ* was not a major contributor.

Since 1978 particularly, many bacons of US origin have been produced using ascorbate or isoascorbate (erythorbate), and these additives can interfere with the determination of nitrite unless adequate precautions are adopted. According to Cassens *et al.* (1974), less than 50% of nitrite added to meat can be determined chemically shortly afterwards by virtue of its reactivity with biological components. Nordin (1969) stated that the rate of depletion of nitrite in a meat product is related exponentially to both pH and temperature. For ham, the rate of decay doubles for about every 12°C increase in temperature and for every 0.86 unit fall in pH. The relationships established by Nordin were independent of the nitrite concentration over the range studied and could be expressed mathematically by the equation: $\log_{10} t_{\frac{1}{2}} = 0.65 - 0.025$ (temperature °C) $+ 0.35$ (pH value), where $t_{\frac{1}{2}}$ is the half-life of nitrite present. Overall, a nitrite loss of around 80% between the processing of meat products and their consumption has been suggested by Buege *et al.* (1978), as compared with that of 85–98% reported by Cassens *et al.* (1979). In 1981, Birdsall estimated the mean nitrite concentration of cured meats at the time of consumption to be 7 mg/kg but this value depends in large measure on the shelf lives of the different meat products. For non-canned items, he accepted measurements made at 15–28 days after processing. This period was extended to 29 days and above for refrigerated canned products and to 60 days plus for shelf stable canned products.

5.4.6 Nitrate and nitrite in milk and milk products

Early studies such as those of Davis and MacDonald (1953) and Hänni (1953) suggested that milk contains very low levels of nitrate of the order of 1 mg NO_3^-/l or less. However, Wright and Davison (1964) reported that the milk from cows used as controls within a feeding experiment contained 4.8 mg nitrate per litre. More recent studies in Denmark (Statens Levnetsmiddelinstitut, 1981) estimated an even higher average concentration of 8 mg/l. If these values are truly representative of milk samples in general they may reflect an increase in food and forage available to the ruminants. Certainly, a fourfold increase in the nitrate concentration of milk to about 20 mg/l was noted in a case in which its intake by cows was particularly high (Wright and Davison, 1964). Furthermore, nitrate levels in dried milk increase when direct-fired processes are employed (Manning *et al.* 1968, Rammell and Joerin 1972). The temperatures of combustion of natural gases currently available in many countries are sufficiently high for the combination of nitrogen and oxygen to produce quantities of oxides of nitrogen capable of producing nitrate and nitrite in contact with milk. Unlike other fuels, natural gases do not usually contain sulphur compounds, which would form sulphur dioxide which can decompose oxides of nitrogen.

Nitrate is not used in cheeses manufactured in the UK but has been used in cheesemaking in continental Europe for about 150 years (MAFF 1987). Nitrate is added to control the production of gas and undesirable flavours by bacteria in a range of cheeses, many of which are imported into the UK. Since microbiological action will probably be involved, the formation of nitrite may occur. Currently regulations (Anon 1979) allow a combined concentration of sodium nitrate and nitrite of 50 mg/kg, of which not more than 5 mg/kg may be of the latter form in cheese other than Cheddar, Cheshire, Grana-padano, Provolone-type cheeses or soft cheese (MAFF 1987).

However, Danish cheeses contain comparable levels of nitrate and nitrite (approximately 10 mg/kg and 0.2 mg/kg respectively) whether or not nitrate had been used in processing (Statens LevnetsMiddelinstitut 1981). Nevertheless, higher nitrate concentrations have been reported in some cheeses (Pedersen *et al.* 1980; Woerner and Fricker, 1960), up to 28 mg/kg. Median nitrate contents of 12-day-old Dutch cheeses in 1975 ranged from 22 to 48 mg/kg (MAFF 1978). Edam cheeses on retail sale in the UK in 1980–81 had nitrate contents within the range 3.1–20 mg/kg (mean 7.5 mg/kg) whilst 15 samples of each of three English-type cheeses (Cheddar, Cheshire and Leicester) were relatively low in nitrate levels (less than 1.0–6.1 mg/kg; mean 2.8 mg/kg) (MAFF 1987). The nitrite contents of the English-type cheeses were all less than 0.4 mg/kg and those of the Edam cheeses were in the range of less than 0.4–0.7 mg/kg (mean less than 0.4 mg/kg).

5.4.7 Nitrate and nitrite contents of cereals and bread

Although the accumulation of nitrate is considered to be far less in grains than in other parts of plants such as the stems and leaves (Hanway *et al.* 1963, Wu and McDonald 1976), it is probable that it is subject to many of the same influences. For instance, the nitrate content of the seeds of winter wheat varies with the strain and growth conditions over the range of 0.4–11 mg NO_3^-/kg according to McNamara *et al.* (1971). The direct drying of cereals using internal combustion dryers can lead to increases in their nitrate and nitrite contents (Fornal *et al.* 1975).

Nitrate concentrations in white flour have been reported as 4–14 mg NO_3^-/kg and approximately 1 mg NO_3^-/kg by Wu and McDonald (1976), and Selenka and Brand-Grimm (1976) respectively; the corresponding value to the latter for the nitrite concentration was 1.2 mg NO_2^-/kg. After baking, however, average levels of 13 mg NO_3^-/kg and 3.4 mg NO_2^-/kg were found in white bread whilst darker rye-based breads, including pumpernickel, had mean concentrations of nitrate and nitrite of 20 mg/kg and 4.3 mg/kg respectively. Further values for nitrite contents have been provided by Harada *et al.* (1972), who found flours to contain about 3 mg/kg; in bread crumbs, noodles and macaroni, however, the levels of nitrite ranged from 10 to 16 mg/kg.

5.5 HUMAN EXPOSURE TO NITRATE AND NITRITE FROM FOOD

A number of different methods can be used to ascertain the average exposure of humans to a dietary component. For instance, use can be made of published tables of the mean consumption of dietary items along with their reported compositions so as to compute the average daily intake of nitrate and nitrite.

Another method is to analyse directly duplicate portions of the meals consumed by individuals. This procedure has a number of advantages over that using food-consumption tables. For instance, the analysis can be made at the time of ingestion, and the actual intake of food is measured on each occasion.

In a study by direct assay of meals in Sweden, Jägerstad and colleagues (Jägerstad and Nilsson 1976, Jägerstad *et al.* 1976) found an average intake of nitrate of 50 mg NO_3^-/day, with a range of 26–81 mg/day per capita, including an unspecified contribution from drinking water. The corresponding mean nitrite intake was 3.7 mg NO_2^-/day, with a range of 0.6–7.3 mg/day. On the other hand, Stephany and Schuller (1978) reported a wider range of values in the Netherlands, again using the duplicate-portion procedure. According to them, the mean ingestion of nitrate was 63 mg NO_3^- person/day, with a range of 6.1–270 mg/day, whilst that of nitrite was 2.3 mg NO_2^-/person/day, with a range of 0.7–14 mg/day. The forty-fold and twenty-fold ranges respectively of the concentrations of nitrate and nitrite were ascribed to seasonal variations in the diets.

The Committee on Nitrite and Alternative Curing Agents in Food (US National Academy of Sciences 1981) has made estimates of the average daily intakes of the various categories of food per capita in the USA. These are reproduced in Table 5.2 along with the average concentrations of nitrate and nitrite and their mean total intakes from the different types of food. Upon this basis, cured meats, to which nitrite and sometimes nitrate are added deliberately during processing, contributed 41% of the former but only 1.5% of the latter burden. However, this estimated percentage of the nitrate intake will be decreased in geographical areas in which water supplies are major contributors.

An estimate of nitrate intake from food in the Federal Republic of Germany by Selenka and Brand-Grimm (1976) gave a figure of 49 mg per day without the inclusion of that derived from drinking water, which would vary greatly from region to region. Results on studies in the UK on total diet samples have been published by the Ministry of Agriculture, Fisheries & Food (MAFF 1987). Nine categories of foods were considered, namely cereals, meats, fish, oils and fats, fruits and sugars, root vegetables (mainly potatoes), other vegetables, non-alcoholic beverages and milk. Most levels of nitrate for foods analysed in 1979 were less than 50 mg/kg except for vegetables and, somewhat surprisingly, two samples of fish which probably gave anomalous values. Estimates of the average total dietary intakes of nitrate were 61 mg/person/day (range 24–102 mg/person/day) to which vegetables contribute

Table 5.2 — Estimates[†] of average daily intakes of various types of food, and their contributions to the intakes of nitrate and nitrite

Source	Daily intake	Mean concentration mg/kg		Mean total intake mg/day	
		Nitrate	Nitrite	Nitrate	Nitrite
Cured meats	30	40	10	1.2	0.30
Fresh measts	60	10	1	0.6	0.06
Vegetables	190	340	0.63	65	0.12
Fruits	220	20	0	4.4	0
Cereals and bread	100	12	2.6	1.2	0.26
Milk and milk products	370	0.5	0	0.19	0
	970			72.6	0.74

[†] As made by the Committee on Nitrite and Alternative Curing Agents in Foods (US National Academy of Sciences, 1981)

about 75%. Of this, 8 mg/person/day was ascribed to non-alcoholic beverages, mostly emanating from the water used to prepare the tea drinks included in the studies. Somewhat less than 4% of the total nitrate was derived from cured meats, of which bacon and ham contributed the major part. The report emphasizes that these values provide no information on daily variations of intake by individuals or on the effects of special habits such as vegetarianism.

Only four of the nine food groups (cereals, meats, root vegetables and other vegetables) were reported to contain nitrite at concentrations above the limit of detection. Most values determined were below the limit of detection; thus different estimates of the average intake of nitrite can be calculated, depending on whether it is assumed that such results are equal to the limit of detection (equivalent to 0.4 mg/kg) or to zero. For this reason, the average nitrite intake was calculated to be either 0.87 or 0.32 mg/person/day. Whatever the correct value, root vegetables were by far the greatest contributor to the nitrite intake, followed by other vegetables, meat and fish, between which the differences were little. These results take no account, of course, of nitrite produced *in vivo* in the consumer consequent upon the ingestion of nitrate.

REFERENCES

Anon (1971). *The Preservatives in Food (Amendment) Regulations 1971 Statutory Instrument (1971) No. 882.*

Anon (1979). *The Preservatives in Food Regulations 1979 Statutory Instrument (1979) No. 752* as amended in 1982 by Statutory Instrument (1982) No. 15.

Anon (1980). *HHS News* (US Department of Health and Human Services), 19 August 1980 issue, pp. 80–35, 1–5.

Barker, A. V., Peck, N. H. and MacDonald, G. E. (1971). Nitrate accumulation in vegetables, I Spinach grown in upland soils, *Agronomy J.* **63**, 126–129.

Birdsall, J. J. (1981). Human dietary nitrite intake and results of animal feeding studies of nitrite. Paper presented at the Public Meeting of the Committee on Nitrite and Alternative Curing Agents in Food, 22 Jan, 1981, National Research Council, National Academy of Sciences, Washington DC, USA, pp. 16.

Brown, J. R. and Smith, G. E. (1967). Nitrate accumulation in vegetable crops as influenced by soil fertility practices. Columbia University of Missouri, pp. 43, Res. Bull. No. 920.

Buege, D. R., Lee, M. H. and Cassens, R. G. (1978). Residual nitrite levels in meat products manufactured by Wisconsin meat processors, University of Wisconsin, Madison, Wisconsin, USA. Available as A2983.

Cantliffe, D. J. (1972). Nitrate accumulation in vegetable crops as affected by photoperiod and light duration (beets, radishes, spinach, beans). *J. Amer. Soc. Hort. Sci.* **97**, 414–418.

Carter, J. N. and Bosma, S. M. (1974). Effect of fertiliser and irrigation on nitrate-nitrogen and total nitrogen in potato tubers. *Agronomy J.* **66**, 263–266.

Cassens, R. G., Sebranek, J. G., Kubberod, G. and Woolford, G. (1974) Where does the nitrite go? *Food Prod. Dev.* **8**, 50–56.

Cassens, R. G., Greaser, M. L., Ito, T. and Lee, M. (1979). Reactions of nitrite in meat. *Food Technol. Chicago* **33**, 46–57.

Cavett, J. J. (1962). The microbiology of vacuum-packed sliced bacon. *J. Appl. Bact.* **25**, 282–289.

Christy, M., Brown, J. R. and Smith, G. E. (1973). Nitrate in soils and plants, Science & Technology Guide, University of Missouri, Columbia Extension Division.

Corré, W. J. and Breimer, T. (1979). Nitrate and nitrite in vegetables. Literature Survey No. 39, Centre for Agricultural Publishing and Documentation, Wageningen, The Netherlands, 85 pp.

Davis, J. G. and MacDonald, F. J. (1953). *Richmond's Dairy Chemistry*, 5th edn, Charles Griffin, London, UK, pp. 603.

Delwiche, C. C. (1970). The nitrogen cycle, *Sci. Am.* **223**, 137–146.

Diem, K. and Lentner, C. (1970). Eds Foods: principal nutrients and minerals. In: *Scientific Tables*, 7th edn, pp. 498–515 Ciba-Geigy Ltd, Basel, Switzerland.

Ellen, G. and Schuller, P. L. (1983). In: *Das Nitrosamin-Problem.* Ed. R. Preussmann, Verlag Chemie, Weinheim, Germany, pp. 98 and 116.

Fornal, L., Fornal, J., Soral-Smietana, M. and Strzalkarska, E. (1975). Nitrate and nitrite content in wheat dried in internal-combustion dryers. *Rocz. Nauk. Roln. Ser. G* **71**, 37–44.

Greenberg, R. A. (1977). Nitrosopyrrolidine in United States cured meat products. In: *Proceedings of the Second International Symposium on Nitrite in Meat Products, 1976, Zeist, The Netherlands.* Eds. B. J. Tinbergen and B. Krol, Centre for Agricultural Publishing and Documentation, Wageningen, The Netherlands.

Habben, J. (1973). Quality constituents of carrots *Daucus carota L*, as influenced by nitrogen and potassium fertilization. *Acta Horticulturae* **29**, 295–304.

Hänni,. H. (1953). Über den Nitratnachweis in Milch. *Mitt. Geb. Lebensmitteluntens. Hyg.* **45**, 502–508.
Hanway, J. J., Herrick, J. B., Willrich, T. L., Bennett, F. C. and McCall, J. T. (1963). *The Nitrate Problem*, Special Report No. 34, Cooperative Extension Services in Agriculture and Home Economics, Iowa State University of Science and Technology, Ames, Iowa, USA, pp. 20.
Harada, M., Nakamura, Y. and Tanimura, A. (1972). Studies on nitrosamines in various foods. *J. Food Hyg. Soc. Japan* **13**, 36–40.
Hata, A. and Ogata, K. (1971a). Nitrate content and nitrate reductase activity in carrot plants. Studies on nitrate and nitrite in horticultural products (part 3). *J. Jap. Soc. Food and Nutr.* **24**, 485–488.
Hata, A. and Ogata, K. (1971b). Changes of nitrate and nitrite contents during storage and cooking of potato tubers. Studies on nitrate and nitrite in horticultural products (Part 2), *J. Jap. Soc. Food and Nutr.* **24**, 345–349.
Herrmann, K. (1972). Über den Nitrat und Nitritgehalt des Gemüses, Obstes und Wassers und deren Bedentung für die Ernährung. *Ernsehr. Umsch.* **11**, 398–402.
Huguet, C., Bonafous, M. and Ducailar, G. (1976). Étude sur la présence des nitrates dans les fruits á pépins et á noyau. *Ann. Nutr. Aliment.* **30**, 673–682.
Jackson, W. A., Steel, J. S. and Boswell, V. R. (1967). Nitrates in edible vegetables and vegetable products. *Proc. Am. Soc. Hortic. Sci.* **90**, 349–352.
Jägerstad, M. and Nilsson, R. (1976). Intake of nitrate and nitrite of some Swedish consumers as measured by the duplicate portion techniques. In: *Second International Symposium on Nitrite in Meat Products, 1976, Zeist, The Netherlands*, Eds. B. J. Tinbergen and B. Krol, Centre for Agricultural Publishing and Documentation, Wageningen, The Netherlands, pp. 283–287.
Jägerstad, M., Norden, A. and Nilsson, R. (1976). Dietary intake of nitrate and nitrite using the duplicate-sampling portion technique. *Ambio* **6**, 276–277.
Kacmar, P. and Bartik, N. (1965). Certain quantitative relations of nitrate and nitrite metabolism in farm animals, IV normal nitrate levels in the organs and body liquids of swine. *Folia Vet.* **9**, 59–63.
Kamm, L., McKeown, G. G. and Smith, D. M. (1965). New colorimetric method for the determination of the nitrate and nitrite content of baby foods. *J. Ass. Off. Anal. Chem.* **48**, 892–897.
Kemp, J. D., Langlois, B. E., Fox, J. D. and Varney, W. Y. (1975). Effects of curing ingredients and holding times and temperatures on organoleptic and microbiological properties of dry-cured sliced ham. *J. Food Sci.* **40**, 634–636.
Kenny, T. A. and Walshe, P. E. (1975). Nitrate and nitrite contents of vegetables and fruit in Ireland. *Ir. J. Agric. Res.* **14**, 349–355.
Kerr, R. H., Marsh, C. T. N., Schroeder, W. F. and Boyer, E. A. (1926). The use of sodium nitrite in the curing of meat. *J. Agric. Res.* **33**, 541–551.
Kilgore, L., Stasch, A. R. and Barrentine, B. F. (1964). Relation of ascorbic acid to nitrate content of turnip greens and to methaemoglobin formation. *Am. J. Clin. Nutr.* **14**, 52–55.
Lee, D. H. K. (1970). Nitrates, nitrites and methaemoglobinaemia, *Environ. Res.* **3**, 484–511.

Li, M., Li, F. and Li, B. (1980). Recent progress in research on esophageal cancer in China. *Adv. Cancer Res.* **33**, 173–249.

Lin, J. K. and Yen, J. Y. (1980). Changes in the nitrate and nitrite contents of fresh vegetables during cultivation and post-harvest storage, *Food Cosmet. Toxicol.* **18**, 597–603.

MAFF (1978). Food Additives and Contaminants, Committee, *Report on the Review of Nitrites and Nitrates in Cured Meats and Cheese*, FAC/REP/27, HMSO.

MAFF (1987). *Nitrate, Nitrite and N-nitroso Compounds in Foods.* 20th Report of the Steering Group on Food Surveillance, Food Surveillance Paper No. 20, HMSO, London.

Manning, P. B., Coulter, S. T. and Jenness, R. (1968), Determination of nitrate and nitrite in milk and dry milk products. *J. Dairy Sci.* **51**, 1725–1730.

Martinoia, E., Heck, V. and Wiemken, A. (1981). Vacuoles as storage compartments for nitrate in barley leaves, *Nature, Lond.* **289**, 292–294.

Massey, R. C., Dennis, M. J., Crews, C., McWeeny, D. J. and Davies, R. (1980). Model system studies on N-nitrosamine formation in relation to cured meat: the non-polar phase and S-nitroso peptides. In: *N-Nitroso Compounds: Analysis, Formation and Occurrence.* Eds E. A. Walker, M. Castegnaro, L. Gricuite and M. Börzsönyi, IARC Scientific Publications No. 31, IARC, Lyon, France, pp. 291–303.

Matsui, H. (1944). On the oxidising substances which appear in Shiozuke extracts of various vegetables. *J. Agric. Chem. Soc., Japan* **16**, 1167–1168.

Maynard, D. N. and Barker, A. V. (1972). Nitrate content of vegetable crops. *Hort. Science* **7**, 224–226.

McNamara, A. S., Klepper, L. A. and Hageman, R. H. (1971). Nitrate content of seeds of certain crop plants, vegetables and weeds. *J. Agric. Food Chem.* **19**, 540–542.

Mills, H. A., Barker, A. V. and Maynard, D. N. (1976). Effects of nitrapyrin on nitrate accumulation in spinach. *J. Am. Soc. Hortic. Sci.* **101**, 202–204.

Nordin, H. R. (1969). The depletion of added sodium nitrite in ham. *Canad. Inst. Food Technol. J.* **2**, 79–85.

Panalaks, T., Iyengar, J. R. and Sen, N. P. (1973). Nitrate, nitrite and dimethylnitrosamine in cured products. *J. Ass. Off. Anal. Chem.* **56**, 621–625.

Panalaks, T., Iyengar, J. R., Donaldson, B. A., Miles, W. F. and Sen, N. P. (1974). Further survey of cured meat products for volatile N-nitrosamines. *J. Ass. Off. Anal. Chem.* **57**, 806–812.

Pedersen, E., Thomsen, J. and Werner, H. (1980). Investigations on formation and occurrence of volatile nitrosamines in Danish cheese. In: *N-Nitroso Compounds: Analysis, Formation and Occurrence.* Eds, E. A. Walker, M. Castegnaro, L. Gricuite and M. Borzsonyi, IARC Scientific Publications No. 31, IARC, Lyon, France, pp. 493–501.

Phillips, W. E. J. (1968a). Changes in the nitrate and nitrite contents of fresh and processed spinach during storage. *J. Agric. Food Chem.* **16**, 88–91.

Phillips, W. E. J. (1968b), Nitrate content of foods — Public Health implications. *Canad. Inst. Food Technol. J.* **1**, 98–103.

Phillips, W. E. J. (1971). Naturally occurring nitrate and nitrite in foods in relation to infant methaemoglobinaemia. *Food Cosmet. Toxicol.* **9**, 219–228.

Pickston, L., Smith, J. M. and Todd, M. (1980). Nitrate and nitrite levels in fruit and vegetables in New Zealand, *Food Technology in New Zealand,* February issue, pp. 11–17.

Rammell, C. G. and Joerin, M. M. (1972).Determination of nitrite in cheese and other dairy products. *J. Dairy Res.* **39**, 89–94.

Richardson, W. D. (1907). The occurrence of nitrates in vegetable foods, in cured meats and elsewhere, *J. Am. Chem. Soc.* **29**, 1757–1767.

Schuphan, W. (1974). The significance of nitrates in food and potable waters. *Qualitas Plantarum — Plant Foods for Human Nutrition* **24**, 19–36.

Selenka, F. and Brand-Grimm, D. (1976). Nitrate and nitrite in human food — calculation of the daily intake and estimation of its range. *Zbl. Bakt. Hyg., I. Abt. Orig. B.* **162**, 449–466.

Sen, N. P., Iyengar, J. R., Donaldson, B. A. and Panalaks, T. (1974). Effect of sodium nitrite concentration on the formation of nitrosopyrrolidine and dimethylnitrosamine in fried bacon. *J. Agric. Food Chem.* **22**, 540–541.

Sen, N. P., Donaldson, B. A. and Charbonneau, C. (1975). Formation of nitrosodimethylamine from the interaction of certain pesticides and nitrite. In: *N-Nitroso Compounds in the Environment.* Eds. P. Bogovski and E. A. Walker, IARC Scientific Publications No. 9, IARC, Lyon, France, pp. 75–79.

Sen, N. P., Donaldson, B. A., Seaman, S., Collins, B. and Iyengar, J. Y. (1977). Recent nitrosamine analysis in cooked bacon. *J. Inst. Can. Sci. Technol. Aliment.* **10**, A13–A15.

Siciliano, J., Krulick, S., Heisler, G., Schwartz, J. H. and White, J. W. (1975). Nitrate and nitrite content of some fresh and processed market vegetables. *J. Agric. Food Chem.* **23**, 461–464.

Statens Levnetsmiddelinstitut (1981). *Nitrat og Nitrit i Kdvarer.* Rapport fra in Intern Arbejdsgrappe, Soborg, Denmark, pp. 63.

Stephany, R. W. and Schuller, P. L. (1978). The intake of nitrate, nitrite and volatile N-nitrosamines and the occurrence of volatile N-nitrosamines in human urine and veal calves. In: *Environmental Aspects of N-Nitroso Compounds.* Eds E. A. Walker, M. Castegnaro, L. Gricuite and R. E. Lyle, IARC Scientific Publications No. 19, IARC, Lyon, France, pp. 443–460.

Usher, C. D. and Telling, G. M. (1975).Analysis of nitrate and nitrite in foodstuffs — critical review, *J. Sci. Food Agric.* **26**, 1793–1805.

US National Academy of Sciences (1981). *The Health Effects of Nitrate, Nitrite and N-nitroso Compounds.* National Academy Press, Washington DC, USA, Chapter 5.

Vulsteke, G. and Biston, R. (1978). Factors affecting nitrate content in field-grown vegetables, *Qual. Plant-Pl. Fds. Hum. Nutr.* **28**, 71–87.

Walters, C. L. and Taylor, A. McM. (1964). Nitrite metabolism by muscle *in vitro. Biochim. Biophys. Acta* **86**, 448–458.

Walters, C. L., Casselden, R. J. and Taylor, A. McM. (1967). Nitrite metabolism by skeletal muscle mitochondria in relation to haem pigments. *Biochim. Biophys. Acta* **143**, 310–318.

Walters, C. L., Edwards, M. W., Elsey, T. S. and Martin, M. (1976). The effect of antioxidants on the production of volatile nitrosamines during the frying of bacon. *Z. für Lebensm. Unters. u. Forsch.* **162**, 377–385.

Ward, F. W., Coates, M. E. and Walker, R. (1986), Nitrate reduction, gastrointestinal pH and N-nitrosation in gnobiotic and conventional rats. *Food Chem. Toxicol.* **24**, 17–22.

White, J. W. (1975). Relative significance of dietary sources of nitrate and nitrite. *J. Agric. Food Chem.* **23**, 886–891.

White, J. W. (1976), Correction: relative significance of dietary sources of nitrate and nitrite. *J. Agric. Food Chem.* **24**, 202.

WHO (1977). *Nitrates, Nitrites and N-nitroso Compounds*, Health Criteria 5, World Health Organisation, Geneva, Switzerland.

Woerner, F. and Fricker, A. (1960). Die Nitrat- und Nitrit-bestimmung in Käse verschiedenen Alters. *Deutsch. Molk Ztg.* **81**, 1345–1348.

Wright, M. J. and Davison, K. L. (1964) Nitrate accumulation on crops and nitrate poisoning in animals. *Adv. Agron.* **16**, 197–247.

Wu, K. Y. and McDonald, C. E. (1976). Effect of nitrogen fertiliser on nitrogen fractions of wheat and flour. *Cereal Chem.* 53 242–249.

Yanagihara, Y., Komoda, H., Yoneyama, H. and Yamada, M. (1963). Nitrate content in juice of various vegetables. *J. Food Hyg. Soc. Japan* **4**, 343–347.

6

Nitrite and nitrate as food additives: rationale and mode of action

T. A. Roberts and R. H. Dainty

6.1 BACKGROUND

Improvements in animal health have greatly reduced the transmission of several diseases once common in man, e.g. bovine tuberculosis, brucellosis. However, apparently healthy animals can carry low numbers of microbes that are subsequently identified as the cause of food-borne illness when foods are inadequately processed or stored under unsuitable conditions for maintaining safety.

Despite advances in animal husbandry and production, improved disease control, and expenditure to improve hygiene at slaughter, there is no evidence that the microbial contamination on carcasses is lower, or that the incidence of bacteria associated with food-borne illness has been reduced. In many countries, food-borne illness has increased greatly over the last 5–10 years.

Consequently, red meat and poultry are recognized as important sources of *Salmonella* spp and *Clostridium perfringens*, both of which are frequent causes of food-borne disease, and *Listeria monocytogenes*. They are also occasional sources of *Staphylococcus aureus, Campylobacter jejuni, Yersinia enterocolitica,* a range of other Enterobacteriaceae, and *Clostridium botulinum* (ICMSF 1980). The factors determining their occurrence are still not well understood, but even with the best husbandry practices, and the most hygienic slaughter conditions, some contamination of the carcass occurs. Because there are no means known to prevent contamination of meat by those microbes, food processors must expect them to be present from time to time and design food processes either to inactivate them, or to prevent their growth.

To be able to grow, bacteria need water and appropriate pH values and temperatures. As the water activity (aw), the pH value or the temperature is reduced, their ability to grow is impaired. Many foods are made safe to eat by a defined heat process and because one or more of these factors is controlled to prevent multiplication of pathogenic bacteria. If a solute such as sodium chloride or sucrose is added, less water is 'available' for growth because more of it is bound to the introduced solute molecules. For example, the aw of fresh meat is *ca.* 0.99, but if it is reduced to *ca.* 0.96, the typical spoilage of chilled fresh (uncured) meats caused by psychrotrophic Gram-negative bacteria is prevented. Reducing the aw to 0.92 or lower prevents the growth of putrefactive anaerobes and mesophilic Enterobacteriaceae even when stored at ambient temperatures. This aw (0.92) conveniently distinguishes between high- and low-aw cured meat products.

Although *Salmonella* spp and *C. perfringens* are common causes of food-borne illness associated with meats, outbreaks are usually the result of temperature abuse (e.g. inadequate cooking followed by unchilled storage), or cross-contamination after cooking, followed by storage with inadequate refrigeration.

Staphylococcus aureus can be a particular problem on unrefrigerated processed meats because it is tolerant of preservatives and can grow at a water activity as low as *ca.* 0.86. The source of *S. aureus* is often man rather than the animal, and control is effected by refrigeration below *ca.* 8°C.

Of the emerging pathogens, *Campylobacter jejuni* is commonly present in the intestinal flora of healthy animals used for food. It has been detected infrequently on beef and lamb carcasses, but more commonly on freshly slaughtered pig and poultry carcasses. Although it appears to fail to survive refrigeration, it has been detected in relatively large numbers on meats at retail. The route by which human campylobacteriosis occurs is unclear.

Many pathogens found on meats are mesophiles and hence unable to grow on meats stored at 0–5°C. Two important exceptions are *Yersinia enterocolitica* and *Listeria monocytogenes,* both of which are psychrotrophic. *Yersinia enterocolitica* is particularly associated with pigs. *Listeria monocytogenes* is present in the intestinal tract of healthy animals and is distributed so widely in the agricultural environment that it is difficult to imagine how it might be eliminated completely. Strict attention to stringent hygienic precautions can reduce its occurrence. Both *L. monocytogenes* and *Y. enterocolitica* have been isolated from a range of raw or mildly processed foods, and there is concern that they may be responsible for more human infection than is currently apparent from epidemiological statistics.

6.2 CURING OF MEATS AND MEAT PRODUCTS

Preservation of meat by drying has been practised for thousands of years. Drying can be achieved by direct removal of water, or indirectly by the addition of salt. The latter became known as curing and led to the variety of cured processed meats now available. The characteristic colour change to pink is the result of conversion of myoglobin to nitric oxide myoglobin (nitrosomyoglobin), and occurs if a source of nitric oxide is provided.

Traditionally, 'saltpetre' was rubbed on the meat surface, when it was realized that its presence as an impurity in sodium chloride used for salting meat caused a cured meat colour to develop. During traditional curing, nitrate is converted to nitrite by bacterial metabolism, but sodium nitrite has now largely replaced nitrate in meat curing, affording better control of the curing process and levels of nitrite to be reduced.

Curing may be combined with other processes including drying, heating, smoking and fermentation. The term 'cured meats' now includes a wide variety of products prepared under different conditions. The essential microbiological characteristic of a cured meat is that it does not putrefy.

Cured-meat products differ greatly in composition and intended eating quality, but the types of bacteria growing on and in them are similar because the main factors controlling their growth are the same in a wide range of products. Over many years their record of bacteriological safety is excellent (Tompkin 1980).

The history of curing was reviewed by Binkerd and Kolari (1975), and the changes in production and marketing that had consequences for the shelf-life and safety of the products by Cerveny (1980). The increased use of refrigeration, the direct application of nitrite, and the pumping of curing salts all improved product stability. Canned cured products, and cured products packaged in vacuum or modified atmospheres, are now common and enjoy an excellent record of stability and safety, although when they were first introduced there were some reservations whether they would be bacteriologically safe. Phosphates are incorporated to minimize shrinkage, reduce cook-out losses, and improve texture with respect to slicing. Ascorbate (or isoascorbate) was included primarily to accelerate the formation of the cured-meat pigment and improve colour stability. Both phosphates (Tompkin 1984) and ascorbates/isoascorbates (Tompkin *et al.* 1979, Robinson *et al.* 1982) are now recognized as important with respect to bacteriological stability.

6.2.1 Antimicrobial aspects of curing

Nitrite is antimicrobial, conferring considerable bacteriological stability to the product, its activity increasing with decreasing pH. It also contributes to the cured meat flavour, which differs from the taste of salt pork. Redox potential (Eh) also affects the activity of nitrite against some bacteria, anaerobic conditions (decreased Eh) increasing its inhibitory effect. Bacterial growth on cured meats is also affected by the chemical composition of the cure (added carbohydrates, phosphates, etc.), the nature of any smoking and cooking processes, and the temperature and gas atmosphere during storage. Whether these factors interact to control microbial growth, or merely act in combination, is difficult to interpret. Much of the microbial growth response can be explained by the pH and aw of the product, the concentration of nitrite present and the temperature of storage.

Lactic acid bacteria are a major component of the microbial flora of most types of cured meat, being tolerant of relatively low aw, e.g. 0.90, and of nitrite. Because they are fermentative, their growth is not restricted by packaging and the common addition of fermentable carbohydrates results in a nutritionally favourable environment for their growth. A flora dominated by lactic acid bacteria is considered normal, and even desirable, for most cured meats. When they cause spoilage it is most commonly due to souring, although other more specific types of spoilage may occur as a result of their growth or metabolism (Egan and Roberts 1987).

The wish to reduce salt consumption has led to milder (lower salt) cures while scrutiny of preservative usage has led to lower nitrite levels. Many products with reduced levels of salt and nitrite store well only under refrigeration. Where refrigeration is not so readily available, heavily salted cured products which are stable at ambient temperatures are more common.

Cured meats can be divided into three categories based upon their heat treatment: raw (unheated), pasteurized (heated to 65–75°C) and severely heated (100–120°C). The microbiology, storage life and safety of these categories of products are discussed in several detailed reviews (Egan and Roberts 1987, Gardner 1983, Lucke 1984).

6.2.2 Role of nitrite in curing of meats

Intensive interest in the role of nitrite in cured meats has resulted from a wish to reduce the level of nitrite in the diet aiming to minimize the risks of consuming preformed nitrosamines and the formation of nitrosamines from ingested nitrites and secondary or tertiary amines in foods.

The relative importance of cured meats as a source of dietary nitrite should be kept in perspective since there are several other sources likely to contribute significantly towards the daily intake, including certain fresh vegetables, dehydrated foods, cigarette smoke, polluted air and endogenous sources such as saliva, intestinal contents and infections of the stomach and bladder.

Nitrite gives cured meat its characteristic colour and flavour, and is important in the control of bacteria, particularly *Clostridium botulinum*. Its importance in meat curing has been the subject of numerous reviews (Krol and Tinbergen 1974, Binkerd and Kolari 1975, Ingram 1976, Tinbergen and Krol 1977, Sofos *et al.* 1979a, Benedict 1980, Hannan 1981, Gray *et al.* 1981, Roberts *et al.* 1981a). Any changes in the regulations reducing permitted levels of nitrite in cured meats are likely to affect the microbiological stability of the product.

It is generally accepted that approximately 40 fg/g nitrite is sufficient to attain and maintain a stable cure colour (Hustad *et al.* 1973) but considerably more nitrite has been regarded as necessary to inhibit the growth of *C. botulinum*, that amount being dependent on a combination of factors including the salt level, initial pH of the product, temperature of storage presence of other chemical additives, the heat process and initial level of microbial contamination.

The wish to minimize the amount of nitrite and/or nitrate added to cured meats resulted in extensive research of the factors controlling the growth of *C. botulinum* in pasteurized cured meats both in the USA (Pivnick *et al.* 1969, Tompkin 1983, Christiansen *et al.* 1973, 1978, Tompkin *et al.* 1978a,b,c,d, Ivey and Robach 1978, Sofos *et al.* 1979a,b,c,d,e, 1980a,b) and in the UK (Roberts *et al.* 1981a,b,c, 1982, Gibson *et al.* 1982, Robinson, *et al.* 1982; Roberts and Gibson 1986). An authoritative overview was constructed by NAS/NRC (1981, 1982).

6.2.3 Regulations on nitrate and nitrite usage in curing

The prime intention of regulations controlling nitrite usage is to prevent undesirably large amounts being consumed. Some countries control the amount of nitrite added to a product at manufacture — e.g. in several EC countries, nitrite may only be added to a meat product as part of a commercially prepared 'curing salt'. In some countries, nitrate is only permitted for specific products — e.g. in Germany the combined addition of nitrite and nitrate is only permitted for large raw hams and some fermented sausages with long ripening periods. In addition to controlling the amount of nitrite added to a product the regulations also state that the residual nitrite must not exceed 100 fg/g (150 fg/g in large raw hams) (Leistner 1981).

Legislation in the UK seeks to control the use of nitrite in food manufacture by monitoring the residual level of nitrite in the product. Until recently all foods in the UK were governed by a single regulation (Preservatives in Foods Regulations 1979) which stated only the maximum permitted residual level of nitrite (200 fg/g) and nitrate (500 fg/g) in the product without specifying when the examination should be made. The regulations remained unaltered since the Preservatives in Foods Regulations (1975) through two amendments (Preservatives in Foods (Amendment) Regulations 1977, 1980). Those regulations did not take into account the levels of other additives such as salt, polyphosphate, ascorbate/isoascorbate, the type of heat process applied, or the temperature of subsequent storage of the product, all of which are relevant to the control of *C. botulinum.* Interactions between several of these factors have been demonstrated (Baird-Parker and Freame 1967, Riemann 1963, Riemann, *et al.* 1972, Roberts and Ingram 1973, Roberts *et al.* 1981a,b,c, 1982, Gibson *et al.* 1982, Robinson *et al.* 1982). Consequently, in respect of control of microbial growth, no factor should be considered in isolation.

In the UK, levels of nitrite and nitrate were reviewed (Food Additives and Contaminants Committee 1978) and a recommendation made to divide cured meats into four categories, each with different permitted nitrite/nitrate levels, according to the heat process given and the type of packaging, making allowance for hermetically sealed containers. No account was taken of level of other curing salts/additives and subsequent storage conditions. Further amendments to the recommendations of the Food Additives and Contaminants Committee (1978) were made in 1982 (Preservatives in Foods (Amendment) Regulations 1982), and the following categories were listed:

1. Cured meat (including cured-meat products) packed in a sterile pack, whether or not it has been removed from the pack. A maximum of 150 fg/g nitrate and nitrite, of which not more than 50 fg/g should be nitrite, expressed in both cases as sodium nitrite.
2. Acidified and/or fermented cured-meat products (including salami and similar products) not packed in a sterile pack. A maximum of 400 fg/g nitrate and nitrite, of which not more than 50 fg/g may be nitrite, expressed in both cases as sodium nitrite.
3. Uncooked bacon and ham; cooked bacon and ham that is not, and has not been, packed in any hermetically sealed container. A maximum of 500 fg/g nitrate and nitrite, of which not more than 200 fg/g may be nitrite, expressed in both cases as sodium nitrite.
4. Any cured meat or cured-meat product not specified above. A maximum of 250 fg/g nitrate and nitrite, of which not more than 150 fg/g may be nitrite, expressed in both cases as sodium nitrite.

6.2.4 Monitoring of nitrate and nitrite levels

There is no general agreement whether the nitrite added (at time of formation) or the residual nitrite (post process) reflects more closely the ability of the product to control growth of *C. botulinum* (Christiansen *et al.* 1973, Hustad *et al.* 1973, Pivnick and Chang 1974, Johnston *et al.* 1969; Christiansen *et al.* 1978, Tompkin *et al.* 1978d) (summarized by Roberts *et al.* 1981a). The 'time to zero residual nitrite', which was literally the time to <10 fg/g (Gibson, *et al.* 1984) proved useful in illustrating the effect of heat treatment, storage temperature and chemical additives (isoascorbate/nitrate/polyphosphate) on residual nitrite in experimental pasteurized cured meats.

Several general conclusions could be drawn. In 'low' pH slurries the rate of loss of nitrite fell with reducing storage temperature, and, although less residual nitrite remained after the higher heat treatments, the rate of loss of that nitrite during storage slowed with increasing heat treatment. If nitrate (500 fg/g) had been added, residual nitrite levels remained higher for longer, especially at initially low-input nitrite levels (100 fg/g). This was more obvious as the heat treatment increased and at low storage temperatures (20°C or below). At 35°C storage, the addition of nitrate had little effect on residual nitrite levels. In some instances residual nitrite increased during storage, particularly after the higher pasteurization treatments and when storage temperature was low. Nitrate-reducing organisms were isolated from many packs but there was insufficient data to relate rise in residual nitrite levels to their presence. Addition of isoascorbate in the absence of nitrate caused residual nitrite levels to fall rapidly and to become undetectable within 20 days irrespective of initial heat treatment, or subsequent storage temperature. If both isoascorbate and nitrate were present, residual nitrite fell at rates similar to those when isoascorbate was present and nitrate absent. Polyphosphate had relatively little effect on the rate of loss of nitrite.

Once in the product, the detectable level of nitrite declines with time during storage. The monitoring of nitrite in the product would detect the serious accidental use, but is only of limited value unless the residual levels can be related to the levels used initially. It is well known that the rate of loss of nitrite is dependent on a number of factors including the initial heat process, the pH of the product and the temperature of storage (Nordin 1969). Hence it would only be possible to estimate the input level from the residual level if all those factors were known. A company monitoring its own product during wholesale storage and distribution or retailing might have such information, but is rarely, if ever, the case with imported products.

From our data, the detection of low levels of nitrite would give no clue whether a product was recently manufactured with an initially low level of nitrite, or was a product which had been stored for several months at a low temperature with an initially modest level of nitrite, or whether it was a product which contained initially illegally high levels of nitrite and nitrate but contained in addition ascorbate or isoascorbate. Hence, from a regulatory point of view, monitoring residual nitrite is of limited value unless performed soon after manufacture.

6.2.5 Residual nitrite and microbiological safety

A separate question is whether the residual nitrite levels imply some degree of microbiological safety. The fate of nitrite in meat products is still not fully explained, although the loss of much of the added nitrite can be accounted for (Cassens *et al.* 1977, 1979, Olsman 1974, 1977). Products containing more nitrite initially are less likely to support the growth of *C. botulinum* (Christiansen *et al.* 1973, Hustad *et al.* 1973, Roberts *et al.* 1981a,b,c) but there has been disagreement for many years among microbiologists whether the inherent stability and microbiological safety of a product is better reflected by the input or residual nitrite level. Elegant experiments by Pivnick and Chang (1974) showed that the residual nitrite level was not the key antimicrobial fraction, but that the nitrite which was lost to the meat system was more likely to be inhibitory to microbes. Attempting to explain their results in a pork homogenate, Christiansen *et al.* (1978) suggested that *C. botulinum* spores germinated irregularly through the storage period while residual nitrite levels fell with time. It was claimed that spores which germinated while detectable nitrite remained died, while if a sufficient number germinated when no residual nitrite remained, growth and toxin production resulted. However, this explanation seems unlikely to apply to experimental cured products containing isoascorbate, which contained no residual (demonstrable) nitrite for much of the 6 months storage,but toxin production did not take place. On the other hand, an identical experimental product, except that it contained no isoascorbate, contained residual nitrite yet supported growth and toxin production by *C. botulinum.*

Separate germination studies on spores of *C. botulinum* types A and B in the absence of salt and nitrite showed only 14% of 226 single-spore isolates germinated and grew within 2 weeks and 36% within 2 months at 35°C storage (Roberts and Thomas 1982). It therefore seems unlikely that all the spores inoculated into our slurries germinated and were subsequently killed within the 10 days during which nitrite levels fell to 11 fg/g and probable that ungerminated spores remained in the absence of nitrite during storage for up to 6 months. Therefore the suggestion offered by Christiansen *et al.* (1978) seems not to be the whole explanation.

Substantial efforts have been made to quantify the relative contributions of the numerous factors that combine to give microbiological safety (Roberts and Ingram 1973, Hauschild 1982, Hauschild *et al.* 1982), and novel approaches proposed that should rapidly lead to a fuller understanding, probably based on mathematical modelling (Gibson *et al.* 1988, Roberts 1989).

6.3 MECHANISM(S) OF MICROBIAL INHIBITION BY NITRITE

In addition to its correct meaning (i.e. the anion NO_2), the term 'nitrite' will be used in this section as a general term to cover the sodium and potassium salts commonly used as food preservatives and the undissociated acid (HNO_2), to which has been attributed an important role in the inhibitory mechanism. Where necessary it will be made clear when nitrite applies specifically to the anion.

Reports of the antimicrobial properties of nitrite appeared in the literature at least as early as the 1920s and 1930s (see Tarr (1941), Shank *et al.* (1962)). A vast literature has subsequently accumulated describing the wide range of organisms whose growth it affects, their relative sensitivities (Tarr 1941), and how these may be markedly influenced by medium composition, the presence of other substances, some inhibitory in their own right (e.g. NaCl), others not (e.g. ascorbate), and other reducing compounds (Castellani and Niven 1955, Fox *et al.* 1981), growth conditions (e.g. pH and O_2 tension), and whether or not nitrite and/or other chemicals (e.g. glucose) (Castellani and Niven 1955) are heated in the medium. At the same time there has been a less voluminous and rather intermittent appearance of papers describing the effects of nitrite on various growth parameters such as lag phase, rate of growth and cell yield (Buchanan and Solberg 1978, Stouthamer 1979), and the biochemical/chemical reactions responsible for the inhibitory effects observed. It is the latter aspects which will now be considered, including the inhibitory effects of nitric oxide (NO) which is readily formed from nitrite in chemical (Castellani and Niven 1955, Williams 1988) and microbiological (e.g. Woods *et al.* 1981) systems.

6.3.1 Action on aerobes

As early as 1927, Warburg is quoted by Shank *et al.* (1962) to have demonstrated reversible and irreversible inhibitory effects of nitric oxide (NO) on respiratory enzyme systems in yeast. In a later study of the effects of various inorganic salts on respiration (O_2 uptake) in *Bacillus cereus,* nitrite was shown to be inhibitory in low concentrations (Ingram 1939), and speculation that nitrite acted by combining with respiratory haem components (i.e. cytochromes) began. Doubt that this could be the complete explanation soon arose from findings with certain spoilage microbes from fish (see Tarr (1941)). Concentrations of nitrite inhibiting these organisms did not significantly alter their aerobic respiration. It was also already known that nitrite could inhibit anaerobes, which do not in general contain haem pigments. Strong evidence supporting Tarr's (1941) contention came some 30 years later in studies of the mechanism of action of a whole range of lipophilic acids used as food additives (Freese *et al.* 1973). In common with many of these acids, particular concentrations of nitrite inhibited cell growth and lowered O_2 uptake and the ATP content of washed cells of *Bacillus subtilis* to a similar extent, but did not inhibit oxidation of either NADH or 3-glycerophosphate by membrane vesicles prepared from the same cultures. These oxidations are cytochrome-dependent, and the authors postulated that the observed inhibition of O_2 uptake and hence lowering of ATP content probably represented a deficiency of oxidizable substrates inside the cells.

Inhibition of serine transport into membrane vesicles by nitrite to an equivalent degree as O_2 uptake in whole cells supported this contention. However, because inhibition of cell growth was pH-dependent but serine uptake was not, the authors concluded that inhibition of transport could not be the sole mechanism of growth inhibition in *B. subtilis.* If inhibition of respiration requires conversion of the nitrite anion to NO, as postulated for the inhibition of non-haem iron systems in clostridia (Woods *et al.* 1981), and is dependent on non-membrane-bound enzymes, then the authors' conclusion would be negated.

This apparent 'uncoupling'of amino acid transport and phosphorylation from the electron transport chain by nitrite received support from the studies of respiratory coupling efficiency in *Paracoccus denitrificans* grown anaerobically under denitrifying conditions (Meijer *et al.* 1979). These authors observed reduced proton-to-oxygen (H^+/O) ratios and increased proton resorption rates during pulsed oxidant assays in the presence of low levels of nitrite. These are precisely the effects observed with classical 'uncouplers' such as carbonyl cyanide-*m*-chlorophenyl hydrazone (CCCP), which act as proton ionophores. Attempts to confirm these findings with a different strain of the organism and to extend them to *Pseudomonas* spp were unsuccessful whether the organisms were grown under anaerobic denitrifying conditions or aerobically (Rake and Eagan 1980). In all three organisms tested, i.e, *P. denitrificans, Ps. aeruginosa* and *Ps.putida,* nitrite significantly reduced H^+/O ratios in washed cell suspensions pulsed with O_2, but rather than being accompanied by increased proton permeability, the opposite was observed, i.e. a decreased proton permeability. In the case of *P. denitrificans,* nitrite did not increase the equilibration time of protons across the membranes of intact cells exposed to pulses of HCl, which CCCP did.

The authors did, however, conclude that their previous observations of nitrite inhibition of O_2 uptake, electron-transport-associated ATP synthesis and active transport of glucose and proline in *Ps. aeruginosa* (Rowe *et al.* 1979) and of active transport of proline in *Escherichia coli* (Yarbrough *et al.* 1980) were consistent with the observed decreases in H^+/O ratios. Active substrate transfer and electron-transport-associated ATP synthesis are dependent on the formation and maintenance of transmembrane proton gradients. Transport processes not dependent on proton gradients, e.g. the group translocation-phosphoenolpyruvate-phosphotransferase-dependent transport of glucose in *E. coli* and *Streptococcus faecalis,* were not inhibited by nitrite (Yarbrough *et al.* 1980), adding more weight to the importance of inhibition of active transport process. The actual mechanism by which the proton gradient is dissipated is not known. Because inhibition of the active transport processes and respiration appeared to occur co-ordinately in *Ps. aeruginosa* and *E.coli,* the authors had earlier speculated that nitrite inhibited the cytochromes of the electron transport chain, thus preventing maintenance of the proton gradient via extrusion (Yarbrough *et al.* 1980). That this is probably not so is suggested by the fact that nitrite inhibits proton extrusion at much lower concentrations than it does respiration.

Nevertheless, inhibition of a purified preparation of cytochrome oxidase from *Ps. aeruginosa* by sub-millimolar concentrations of nitrite has been presented (Yang 1985). The inhibition was uncompetitive (cf. NO inhibition of clostridial pyruvate: ferredoxin oxidoreductase and yeast pyruvate decarboxylase (McMindes and Siedler 1988)), and was quite unlike the competitive inhibition of cytochrome oxidase by classical inhibitors such as cyanide and azide. Further evidence that such a mechanism could not be the sole inhibitory mechanism was provided by the authors' findings of greater inhibition of malate- and glucose-dependent O_2 uptake in whole cells than NADH or *N,N,N',N'* tetramethyl-*p*-phenylenediamine/TMP-dependent O_2 uptake.

A more direct effect of nitrite on bacterial cell membrane components has recently been demonstrated by Buchman and Hansen (1987), who associated nitrite inhibition of outgrowth of germinated spores of *B. cereus* with reaction at specific sulphydryl groups. Not only did nitrite protect against the incorporation of the specific sulphydryl reagent iodoacetate into membranes but it also protected against the irreversible loss of viability caused by the associated carboxymethylation of the sulphydryl groups. Washing nitrite-inhibited cells with water reversed the inhibition and also eliminated the protective effects against iodoacetate. It is perhaps significant that nitrite inhibition of active transport of glucose and proline in *Ps. aeruginosa* is also reversible by washing with water (Rowe *et al.* 1979). Although the specific sulphydryl groups involved in nitrite inhibition were not identified, earlier studies with S-nitrosothiols, which are nitrite analogues, and with other sulphydryl agents and nisin, suggest that they are components of, or in contact with, membrane permeases (Hansen and Levin 1975, Morris and Hansen 1981, Morris *et al.* 1984), and that all these inhibitory agents probably act at the same sites.

Evidence therefore seems to be getting stronger that a major site of action of nitrite in aerobically grown cells is bacterial cell membranes and associated processes. For the sake of completeness it should be remembered that other sites of inhibitory action of nitrite have been reported in aerobes including aldolase in *E. coli, P. aeruginosa* and *S. faecalis* (Yarbrough *et al.* 1980) and pyruvate-metabolizing systems in several other organisms. Both *Vibrio comma* (Bernheim 1943) and a *Fusarium* sp (Nord and Mull 1945) accumulated pyruvate when growing on glucose in the presence of nitrite, while *Staphylococcus aureus* produced lower levels of acetoin in favour of acetate in the presence of nitrite (Buchanan and Solberg 1978). There had been earlier speculation that nitrite inhibits *S. aureus* by reacting with the sulphydryl groups of the coenzyme A or lipoic acid cofactors involved in pyruvate metabolism, although no evidence was presented (Buchanan and Solberg 1972). Recently, the pyruvate decarboxylase from brewer's yeast has been shown to be inhibited by nitrite, and even more actively by NO (McMindes and Siedler 1988). This enzyme is not dependent on either lipoic acid or coenzyme A.

6.3.2 Action on anaerobes

Discussion of the effects of nitrite on pyruvate metabolism provides an appropriate starting point for anaerobic systems. Pyruvate metabolism in anaerobes has been almost exclusively the centre of attraction for studies of the inhibitory effects of nitrite.

In addition to the study of pyruvate decarboxylase from brewer's yeast by McMindes and Siedler (1988), those authors also investigated the effects of nitrite and NO on pyruvate:ferredoxin oxidoreductase (PFO) from *Clostridium perfringens*. In the concentrations inhibiting the yeast decarboxylase, nitric oxide, but not nitrite, inhibited PFO, thus confirming earlier findings with *C. sporogenes* (Woods *et al.* 1981) and *C. botulinum* types A, B (proteolytic and non-proteolytic), C, D and E (Woods and Wood 1982). In both of these studies, pyruvate accumulated when nitrite was added to glucose-metabolizing cultures. In *C. sporogenes* the rate of CO_2 and H_2 evolution was also reduced and there was a rapid and marked decline in the ATP content of the cells. These observations provide strong evidence of inhibition of the phosphoroclastic split of pyruvate in anaerobic systems. In cell-free systems of *C. sporogenes* (Woods *et al.* 1981) the pyruvate-dependent formation of acetyl phosphate was not inhibited at the highest nitrite concentrations tested but the pyruvate-dependent reduction of NAD, which is ferredoxin-dependent, and ATP production was affected at quite low concentrations. To try to define the site of action more closely, cell-free extracts of *C. sporogenes* were fractionated into pure ferredoxin and ferredoxin-free protein extracts after growth in the presence and absence of nitrite. Determination of pyruvate-dependent NAD reduction in reconstituted systems, reflecting all possible combinations of these fractions, led the authors to conclude that the PFO enzyme was the main site of inhibitory activity. They also concluded that NO, which was detected in small amounts in extracts incubated with nitrite anion, was the major inhibitory chemical moiety. However, their data show that NO bubbled through the ferredoxin fraction prior to reconstitution caused greater inhibition than when bubbled through the ferredoxin-free extract containing PRO activity. The results of reconstitution experiments also showed significant inhibition (*ca.* 80%) when ferredoxin from nitrite-grown cells was incubated with PFO-containing extract from cells grown without nitrite. Taken together with the observed inhibition of NAD reduction and ATP formation in the presence of nitrite, but not of acetyl phosphate formation (see above), the authors' conclusions do not appear to be a totally accurate and complete reflection of their findings. Whatever the precise details of the inhibitory mechanism ultimately prove to be, there can be little doubt that inhibition of such a central mechanism to provision of the organisms requirements for ATP is likely to be an important inhibitory mechanism. Additionally, any effect of nitrite in preventing the reoxidation of ferredoxin and NAD reduced in pyruvate breakdown, which normally occurs through H_2 formation catalysed by hydrogenase, would have equally serious consequences for cell metabolism and growth.

Subsequent experiments on the *in vivo* and *in vitro* effects of nitrite on the components of the phosphoroclastic system of *C. botulinum* in fact led to the conclusion that the inhibitory effects of nitrite were due, at least in part, to the inactivation of both ferredoxin and pyruvate-ferredoxin oxidoreductase (Carpenter *et al.* 1987). Again, NO gas was more inhibitory that nitrite. Earlier work by these authors, using electron spin resonance spectroscopy (esr), had in fact detected signals characteristic of nitrosyl complexes of iron-containing molecules (Reddy *et al.* 1983). The authors therefore conjectured that reaction of nitrite (NO) with the non-haem iron centres of ferredoxin and PFO was the likely site of inhibition, a theory proposed years earlier by Tompkin *et al.* (1978e). Such non-haem iron sites are characterized by the presence of sulphur atoms, and it is of interest that Fe–S–NO complexes, as in Roussin's black and red salts, have long been conjectured to play a role in the enhanced activity of nitrite in heated systems, particularly in the inhibition of clostridia (Perigo and Roberts 1968, van Roon 1974, Moran *et al.* 1975). Both nitrite and NO have also been shown to inhibit the activity of nitrogenase, another non-haem iron-containing protein, in *C. pasteurianum* (Meyer 1981).

Clearly, more work is needed to clarify the situation completely with regard to which components are the most sensitive and whether the nitrite anion or NO is the more effective molecular species.

As in the case of aerobes it would be wise to remember that there have been reports of nitrite inhibition of other enzymes. A complete loss of glyceraldehyde-3-phosphate dehydrogenase activity and a substantial inhibition of aldolase were observed when growth of *C. perfringens* was inhibited by nitrite (O'Leary and Solberg 1976). Both enzymes are involved in glucose metabolism. In addition there was a very substantial fall in the level of assayable sulphydryl groups and addition of the sulphydryl reducing agent dithiothreitol restored some activity to the inhibited dehydrogenase. A totally novel observation in anaerobically grown *Lactobacillus plantarum* is the accumulation of vastly increased levels of manganese ions in the presence of growth-inhibitory levels of nitrite (Collins-Thompson and Thomson 1986). This was not the case in aerobically grown cells, which were resistant to the same levels of nitrite. The association of this phenomenon with inhibition was not established, if indeed any association exists.

In conclusion, although significant progress has been made in understanding the mechanism of nitrite inhibition of microbial growth, neither the precise metabolic step(s) nor the exact chemical interactions have been determined with certainty. On balance, the evidence presently points to different mechanisms in aerobes and anaerobes, albeit with electron transfer proteins perhaps being important in both types of organism. However, sight should not be lost of the possibility that the inherent reactivity (Williams 1988) of nitrite or NO with the –SH, $-NH_2$ and –OH groups characteristic of so many biological molecules could mean that no single mechanism of inhibition will ultimately be found to be of central importance. For example, despite all the attention paid to pyruvate metabolism in aerobes, Woods *et al.* (1981) showed that glucose transport in *C. sporogenes* was inhibited along with the phosphoroclastic system. Whether the two were linked or independent is not known, but it is possible that the glucose transport could have been a result of inhibition of membrane sulphydryls as in bacilli (Buchman and Hansen 1987).

REFERENCES

Baird-Parker, A. C. and Freame, B. (1967). Combined effect of water activity, pH and temperature on the growth of *Clostridium botulinum* from spore and vegetative cell inocula. *J. Appl. Bacteriology* **30**, 420–429.

Benedict R. C. (1980). Biochemical basis for nitrite-inhibition of *Clostridium botulinum* in cured meat. *J. Food Protection* **43**, 877–891.

Bernheim, F. (1943). The significance of the amino groups for the oxidation of various compounds by *Cholera vibrio* (V. comma). *J. Biological Chemistry* **157**, 221–240.

Binkerd, E. F. and Kolari, O. E. (1975). The history and use of nitrate and nitrite in the curing of meat. *Food and Cosmetics Toxicology* **13**, 655–661.

Buchanan, R. L. and Solberg, M. (1972). Interaction of sodium nitrite, oxygen and pH on growth of *Staphylococcus aureus*. *J. Food Sci.* **37**, 81–85.

Buchanan, R. L. and Solberg, M. (1978). Influence of sodium nitrite on the aerobic catabolism of glucose by *Staphylococcus aureus*. *J. Food Safety* **1**, 189–200.

Buchman III, G. W. and Hansen, J. N. (1987). Modification of membrane sulfhydryl groups in bacteriostatic action of nitrite. *Appl. and Environmental Microbiology* **53**, 79–82.

Carpenter, C. E., Reddy, D. S. A. and Cornforth, D. P. (1987). Inactivation of clostridial ferredoxin and pyruvate-ferredoxin oxidoreductase by sodium nitrite. *Appl. and Environmental Microbiology* **53**, 549–552.

Cassens, R. G., Woolford, G., Lee, S. H. and Goutefongea, R. (1977). Fate of nitrite in meat. In: *Proceedings of the Second International Symposium on*

Nitrite in Meat Products, Zeist, 1976. Eds B. J. Tinbergen and B. Krol, Wageningen, PUDOC, pp. 95–109.

Cassens, R. G., Greaser, M. L., Ito, T. and Lee, M. (1979). Reactions of nitrite in meat. *Food Technol.* **33**, 46–57.

Castellani, A. G. and Niven, Jr, C. F. (1955). Factors affecting the bacteriostatic action of sodium nitrite. *Appl. Microbiology* **3**, 154–159.

Cerveny, J. G. (1980). Effects of changes in the production and marketing of cured meats on the risk of botulism. *Food Technol.* **34**, 240–243. (F3275)

Christiansen, L. N., Johnston, R. W., Kautter, D. A., Howard, J. W. and Aunan, W. J. (1973). Effect of nitrite and nitrate on toxin production by *Clostridium botulinum* and nitrosamine formation in perishable canned comminuted cured meat. *Appl. Microbiology* **25**, 357–367 (*erratum* **26**, 653).

Christiansen, L. N., Tompkin, R. B. and Shaparis, A. B. (1978). Fate of *Clostridium botulinum* in perishable canned cured meat at abuse temperature. *J. Food Protection* **41**, 354–355.

Collins-Thompson, D. L. and Thomson, I. Q. (1986). Changes in manganese content in *Lactobacillus plantarum* during inhibition with sodium nitrite. *J. Food Protection* **49**, 602–604

Egan, A. F. and Roberts, T. A. (1987). Microbiology of meat and meat products. In: *Essays in Agricultural and Food Microbiology.* Eds J. R. Norris and G. Pettipher, Wiley, pp. 167–197.

Food Additives and Contaminants Committee (1978) *Report on the review of nitrites and nitrates in cured meats and cheese.* HMSO Report No. 27.

Fox, J., Fiddler, R. and Wasserman, A. (1981). Initial reaction intermediates in the oxidation of ascorbic acid by nitrous acid. *J. Food Protection* **44**, 28–32.

Freese, E., Sheu, C. W. and Galliers, E. (1973). Function of lipophilic acids as antimicrobial food additives. *Nature* **241**, 321–325.

Gardner, G. A. (1983). Microbial spoilage of cured meats. In: *Food Microbiology: Advances and Prospects.* Eds T. A. Roberts and F. A. Skinner, Society for Applied Bacteriology Symposium Series No. 11, Academic Press. London, pp. 179–202.

Gibson, A. M., Roberts, T. A. and Robinson, A. (1982). Factors controlling the growth of *Clostridium botulinum* types A and B in pasteurized cured meats. IV. The effect of pig breed, cut and batch of pork. *J. Food Technology* **17**, 471–482.

Gibson, A. M., Roberts, T. A. and Robinson, A. (1984). Factors controlling the growth of *Clostridium botulinum* types A and B in pasteurized cured meats VI. Nitrite monitoring during storage of pasteurized pork slurries. *J. Food Technology* **19**, 29–44.

Gibson, A. M., Bratchell, N. and Roberts, T. A. (1988). Predicting microbial growth: growth responses of salmonellae in a laboratory medium as affected by pH, sodium chloride and storage temperature. *Int. J. Food Microbiology* **6**, 155–178.

Gray, J. I., MacDonald, B., Pearson, A. M. and Morton, I. D. (1981). Role of nitrite in cured meat flavour: a review. *J. Food Protection* **44**, 302–312.

Hannan, R. S. (1981). Use of nitrite as a preservative in British meat products. *Proc, Inst. Food Sci. and Technology* **14**, 8–14.

Hansen, J. N. and Levin, R. A. (1975). Effect of some inhibitors derived from nitrite on macromolecular synthesis in *Bacillus cereus*. *Appl. Microbiology* **30**, 862–869.

Hauschild, A. H. W. (1982). Assessment of botulism hazards from cured meat products. *Food Technology* **36**, 95–104.

Hauschild, A. H. W., Hilsheimer, R., Jarvis, G. and Raymond, D. P. (1982). Contribution of nitrite to the control of *Clostridium botulinum* in liver sausage. *J. Food Protection* **45**, 500–506.

Hustad, G. O., Cerveny, J. G., Trenk, H., Deibel, R. H., Kautter, D. A., Fazio, T., Johnston, R. W. and Kolari, O. E. (1973). Effect of sodium nitrite and sodium nitrate on botulinal toxin production and nitrosamine formation in weiners. *Appl. Microbiology* **26**, 22–26.

ICMSF (International Commission on Microbiological Specifications for Foods) (1980). *Microbial Ecology of Foods,* Volume I: *Factors Affecting Life and Death of Microorganisms.* Volume II: *Food Commodities,* Academic Press, New York.

Ingram, M. (1939). The endogenous respiration of *Bacillus cereus* II. The effect of salts on the rate of absorption of oxygen. *J. Bacteriology* **38**, 613–629.

Ingram, M. (1976). The microbiological roles of nitrite in meat products. In: *Microbiology in Agriculture, Fisheries and Food.* F. A. Skinner and J. G. Carr Society for Applied Bacteriology Symposium Series No. 4. Academic Press, London, pp. 1–18.

Ingram, M. and Roberts, T. A. (1971). Application of the 'D-concept' to heat treatments involving curing salts. *J. Food Technology* **6**, 21–28.

Ivey, F. J. and Robach, M. C. (1978). Effect of sorbic acid and sodium nitrite on *Clostridium botulinum* outgrowth and toxin production in canned comminuted pork. *J. Food Sci.* **43**, 1782–1785.

Johnston, M.A., Pivnick, H. and Samson, J. M. (1969). Inhibition of *Clostridium botulinum* by sodium nitrite in a bacteriological medium and in meat. *Canadian Inst. Food Technology J.* **2**, 52–55.

Krol, B. and Tinbergen, B. J. (Eds) (1974). *Proceedings of the International Symposium on Nitrite in Meat Products.* PUDOC, Wageningen, Netherlands.

Leistner, L. (1981). New nitrite regulations of West Germany. *Fleischwirtschaft* **61**, 252–254.

Lucke, F.-K. (1984). Fermented sausages. In: *Microbiology of Food Fermentations.* Ed. B. J. B. Wood, Volume 2, Applied Science Publishers, London, pp. 41-83.

McMindes, M. K. and Siedler, A. J. (1988). Nitrite mode of action: inhibition of yeast pyruvate decarboxylase (E.C.4.1.1.1) and clostridial pyruvate:ferredoxin oxido-reductase (E.C.1.2.7.1) by nitric oxide. *J. Food Sci.* **53**, 917–919, 931.

Meijer, E. M., Van der Zwaan, J. W., Weaver, R. and Stouthamer, A. H. (1979). Aerobic respiration and energy conservation in *Paracoccus denitrificans*. *European J. Biochem.* **96**, 69–76.

Meyer, J. (1981). Comparison of carbon monoxide, nitric oxide and nitrite as inhibitors of the nitrogenase from *Clostridium pasteurianum*. *Arch. Biochem. and Biophys.* **210**, 246–256.

Moran, D. M., Tannenbaum, S. R. and Archer, M. C. (1975). Inhibitor of *Clostridium perfringens* formed by heating sodium nitrite in a chemically defined medium. *Appl. Microbiology* **30**, 838–843.

Morris, S. L. and Hansen, J. N. (1981). Inhibition of *Bacillus cereus* spore outgrowth by covalent modification of a sulfhydryl group by nitrosothiol and iodoacetate. *J. Bacteriology* **148**, 465–471.

Morris, S. L., Walsh, R. C. and Hansen, J. N. (1984). Identification and characterization of some bacterial membrane sulfhydryl groups which are targets of bacteriostatic and antibiotic action. *J. Biological Chemistry* **259**, 13590–13594.

NAS/NRC (National Academy of Sciences/National Research Council). (1981). *The Health Effects of Nitrate, Nitrite and N-nitroso Compounds.* Part 1 of a 2-part study by the Committee on Nitrite and Alternative Curing Agents in Foods. Assembly of Life Sciences. National Academy Press, Washington DC.

NAS/NRC (National Academy of Sciences/National Research Council). (1982). *The Alternatives to the Current Use of Nitrite in Foods.* Part 2 of a 2-part study by the Committee on Nitrite and Alternative Curing Agents in Foods. Assembly of Life Sciences. National Academy Press, Washington, DC.

Nord, F. F. and Mull, R. P. (1945). Recent progress in the biochemistry of fusaria. *Advances in Enzymology* **5**, 165–205.

Nordin, H. R. (1969). The depletion of added sodium nitrite in ham. *Canadian Inst. Food Technology J.* **2**, 79–85.

O'Leary, V. and Solberg, M. (1976). Effect of sodium nitrite inhibition on intracellular thiol groups and on the activity of certain glycolytic enzymes in *Clostridium perfringens. Appl. and Environmental Microbiology* **31**, 208–212.

Olsman, W. J. (1974). About the mechanism of nitrite loss during storage of cooked meat products. In: *Proceedings of the International Symposium on Nitrite in Meat Products, Zeist, 1973* Eds B. Krol, and B. J. Tinbergen, PUDOC, Wageningen, The Netherlands, pp. 129–137.

Olsman, W. J. (1977). Chemical behaviour of nitrite in meat products. I. The stability of protein bound nitrite during storage. In: *Proceedings of the Second International Symposium on Nitrite in Meat Products, Zeist, 1976.* Eds B. J. Tinbergen, and B. Krol, PUDOC, Wageningen, The Netherlands, pp. 101–109.

Perigo, J. A. and Roberts, T. A. (1968). Inhibition of clostridia by nitrite. *J. Food Technology* **3**, 91–94.

Pivnick, H., Barnett, H. W., Nordin, H. R. and Rubin, L. J. (1969). Factors affecting the safety of canned, cured, shelf-stable luncheon meat inoculated with *Clostridium botulinum. Canadian Inst. Food Technology J.* **2**, 141–148.

Pivnick, H. and Chang, P. C. (1974). Perigo effect in pork. In: *Proceedings of the International Symposium on Nitrite in Meat Products, Zeist, 1973.* Eds B. Krol, and B. J. Tinbergen, PUDOC, Wageningen, The Netherlands, pp. 111–116.

Preservatives in Foods Regulations (1975). Statutory Instruments No. 1487 (Food and Drugs, Composition and Labelling). HMSO, London.

Preservatives in Foods (Amendment) Regulations (1977). Statutory Instruments No. 645 (Food and Drugs, Composition and Labelling). HMSO, London.

Preservatives in Foods (Amendment) Regulations (1979). Statutory Instruments No. 752 (Food and Drugs, Composition and Labelling). HMSO, London.

Preservatives in Foods (Amendment) Regulations (1980). Statutory Instruments No. 931 (Food and Drugs, Composition and Labelling). HMSO, London.

Preservatives in Foods (Amendment) Regulations (1982). Statutory Instruments No. 15 (Food Composition and Labelling). HMSO, London.

Rake, J. B. and Eagan, R. G. (1980). Inhibition, but not uncoupling, of respiratory energy coupling of three bacterial species by nitrite. *J. Bacteriology* **144**, 975–982.

Reddy, D., Lancaster, J. R. and Cornforth, D. P. (1983). Nitrite inhibition of *Clostridium botulinum:* electron spin resonance detection of iron–nitric oxide complexes. *Science* **221**, 769–770.

Riemann, H. (1963). Safe heat processing of canned cured meats with regard to bacterial spores. *Food Technology* **17**, 39–42, 45–46, 49.

Riemann, H., Lee, W. H. and Genigeorgis, C. (1972). Control of *Clostridium botulinum* and *Staphylococcus aureus* in semi-preserved meat products. *J. Milk Food Technology* **35**, 514–523.

Roberts, T. A. (1989). Combinations of antimicrobials and processing methods. *Food Technology,* **43**, 156–163.

Roberts, T. A. and Ingram, M. (1973). Inhibition of growth of *Clostridium botulinum* at different pH values by sodium chloride and sodium nitrite. *J. Food Technology* **8**, 467–475.

Roberts, T. A., Gibson, A. M. and Robinson, A. (1981a). Factors controlling the growth of *Clostridium botulinum* types A and B in pasteurized cured meats. I. Growth in pork slurries prepared from 'low' pH meat (pH range 5.5–6.3). *J. Food Technology* **16**, 239–266.

Roberts, T. A., Gibson, A. M. and Robinson, A. (1981b). Factors controlling the growth of *Clostridium botulinum* types A and B in pasteurized cured meats. II. Growth in pork slurries prepared from 'high' pH meat (pH range 6.3–6.8). *J. Food Technology* **16**, 267–281.

Roberts, T. A., Gibson, A. M. and Robinson, A. (1981c). Prediction of toxin production by *Clostridium botulinum* in pasteurized pork slurry. *J. Food Technology* **16**, 337–355.

Roberts, T. A., Gibson, A. M. and Robinson, A. (1982). Factors controlling the growth of *Clostridium botulinum* types A and B in pasteurized cured meats. III. The effect of potassium sorbate. *J. Food Technology* **17**, 307–326.

Roberts, T. A. and Thomas, J. (1982). Germination and outgrowth of single spores of *Clostridium botulinum* and putrefactive anaerobes. *J. Appl. Bacteriology* **53**, 317–321.

Roberts, T. A. and Gibson, A. M. (1986). Chemical methods for controlling *Clostridium botulinum* in processed meats. *Food Technology* **40**, 163–171 and 176.

Robinson, A., Gibson, A. M. and Roberts, T. A. (1982). Factors controlling the growth of *Clostridium botulinum* types A and B in pasteurized meats. V. Prediction of toxin production: non-linear effects of storage temperature and salt concentration. *J. Food Technology* **17**, 727–744.

Rowe, J. J., Yarbrough, J. M., Blake, J. B. and Eagon, R. G. (1979). Nitrite inhibition of aerobic bacteria. *Current Microbiology* **2**, 51–54.

Shank, J. L., Silliker, J. H. and Harper, R. H. (1962). The effect of nitric oxide on bacteria. *Appl. Microbiology* **10**, 185–189.

Sofos, J. N, Busta, F. F. and Allen, C. E. (1979a). Botulism control by nitrite and sorbate in cured meats: a review. *J. Food Protection* **42**, 739–770.

Sofos, J. N., Busta, F. F. and Allen, C. E. (1979b). Sodium nitrite and sorbic acid effects on *Clostridium botulinum* spore germination and total microbial growth in chicken frankfurter emulsions during temperature abuse. *Appl. and Environmental Microbiology* **37**, 1103–1109.

Sofos, J. N., Busta, F. F., Bhothipaksa, K. and Allen, C. E. (1979c). Sodium nitrite and sorbic acid effects on *Clostridium botulinum* toxin formation in chicken frankfurter-type emulsions. *J. Food Sci.* **44**, 668–672.

Sofos, J. N., Busta, F. F. and Allen, C. E. (1979d). *Clostridium botulinum* control by sodium nitrite and sorbic acid in various meat and soy protein formulations. *J. Food Sc.* **44**, 1662–1667, 1671.

Sofos, J. N., Busta, F. F. and Allen, C. E. (1979e). Effect of sodium nitrite on *Clostridium botulinum* toxin production in frankfurter emulsions formulated with meat and soy proteins. *J. Food Science* **44**, 1267–1271.

Sofos, J. N., Busta, F. F. and Allen, C. E. (1980a). Influence of pH on *Clostridium botulinum* control by sodium nitrite and sorbic acid in chicken emulsions. *J. Food Sci.* **45**, 7–12.

Sofos, J. N., Busta, F. F., Bhothipaksa, K., Allen, C. E., Robach, M. C. and Paquette, M. W. (1980b). Effects of various concentrations of sodium nitrite and potassium sorbate on *Clostridium botulinum* toxin production in commercially prepared bacon. *J. Food Sci.* **45**, 1285–1292.

Stouthamer, A. H. (1979). The search for correlation between theoretical and experimental growth yields. *Int. Rev. in Biochemistry* **21**, 1–47.

Tarr, H. L. A (1941). Bacteriostatic action of nitrites. *Nature* **147**, 417–418.

Tinbergen, B. J. and Krol, B. (Eds) (1977). *Proceedings of the Second International Symposium on Nitrite in Meat Products.* PUDOC, Wageningen, Netherlands.

Tompkin, R. B. (1980). Botulism from meat and poultry products — a historical perspective. *Food Technology* **34**, 229–236, 257.

Tompkin, R. B. (1983). Nitrite. In: *Antimicrobials in Foods.* Eds A. L. Branen and P. M. Davidson, Chapter 8, Marcel Dekker, New York, pp. 205–256.

Tompkin, R. B. (1984). Indirect antimicrobial effect in foods: phosphates. *J. Food Safety* **6**, 13–27.

Tompkin, R. B., Christiansen, L. N. and Shaparis, A. B. (1978a). Enhancing nitrite inhibition of *Clostridium botulinum* with isoascorbate in perishable canned cured meat. *Appl. and Environmental Microbiology* **35**, 59–61.

Tompkin, R. B., Christiansen, L. N. and Shaparis, A. B. (1978b). Effect of prior refrigeration on botulinal outgrowth in perishable canned cured meat when temperature abused. *Appl. and Environmental Microbiology* **35**, 863–866.

Tompkin, R. B., Christiansen, L. N. and Shaparis, A. B. (1978c). Antibotulinal role of isoascorbate in cured meat. *J. Food Sci.* **43**, 1368–1370.

Tompkin, R. B., Christiansen, L. N. and Shaparis, A. B. (1978d). Causes of variation in botulinal inhibition in perishable canned cured meat. *Appl. and Environmental Microbiology* **35**, 886–889.

Tompkin, R. B., Christiansen, L. N. and Shaparis, A. B. (1978e). The effect of iron on botulinal inhibition in perishable canned cured meat. *J. Food Technology* **13**, 521–527.

Tompkin, R. B., Christiansen, L. N. and Shaparis, A. B. (1979). Isoascorbate level and botulinal inhibition in perishable canned cured meat. *J. Food Sci.* **44**, 1147–1149.

van Roon, P. S. (1974). Inhibitors in cooked meat products. In: *Proceedings of the International Symposium on Nitrite in Meat Products, Zeist, 1973.* Eds. B. Krol, and B. J. Tinbergen, PUDOC, Wageningen, The Netherlands, pp. 117—124.

Williams, D. L. H. (1988). *Nitrosation.* Cambridge University Press, UK.

Woods, L. F. J., Wood, J.M. and Gibbs, P. A. (1981). The involvement of nitric oxide in the inhibition of the phosphoroclastic system in *Clostridium sporogenes* by sodium nitrite. *J. General Microbiology* **125**, 399—406.

Woods, L. F. J. and Wood, J. M. (1982). A note on the effect of nitrite inhibition on the metabolism of *Clostridium botulinum. J. Appl. Bacteriology* **52**, 109–110.

Yang, T. (1985). Mechanism of nitrite inhibition of cellular respiration in *Pseudomonas aeruginosa. Current Microbiology* **12**, 35–40.

Yarbrough, J. M., Rake, J. B. and Eagon, R. G. (1980). Bacterial inhibitory effects of nitrite: inhibition of active transport, but not of group translocation and of intracellular enzymes. *Appl. and Environmental Microbiology* **39**, 830–834.

7

Human exposure, pharmacology and metabolism of nitrate and nitrite

P. J. Packer and S. A. Leach

7.1 INTRODUCTION

Nitrate and nitrite are fundamental components of the global nitrogen cycle and are therefore found throughout the environment. In health, the major source of human nitrate exposure is from the oral ingestion of the preformed ion (Hartman 1982). The main contributions are from vegetables and potable drinking water (Central Water Planning Unit 1977), with a lesser contribution from cured meats (see other relevant chapters). Although the permitted levels of nitrate and nitrite added to meats have been decreased in recent years (Statutory Instruments 1962, 1979, 1982), other factors have led to a net increase in exposure to nitrate for many populations. For example, increased emissions of nitrogen oxides from fuel combustion and their deposition as a component of acid rain (Brimblecombe and Steadman 1982), increased sewage recycling, and the increased use of nitrate-based fertilizers (White 1983) have all progressively increased environmental nitrate levels. The result of the last-mentioned in particular has been to increase the nitrate content of many vegetables and drinking water supplies (White 1983). As a further source of nitrate exposure, apart from exogenous sources, mammalian endogenous nitrate biosynthesis from L-arginine has also been highlighted in several recent studies (Marletta 1988, Leaf *et al.* 1989, 1990). However, the importance of this nitrate source is probably more important in terms of overall nitrate exposure when the immune system of the host is appropriately stimulated.

Exposure to nitrate and nitrite has been postulated to have several deleterious health effects. Although Hartman (1982) has tentatively suggested some direct health effects of nitrate *itself*, at normal levels of exposure it has no proven toxicological action. The major effects of nitrate appear to be mediated indirectly through its endogenous metabolites, nitrite and N-nitroso compounds, most of which are known to have profound toxicological effects. Ingested levels of nitrite and N-nitroso compounds from exogenous dietary sources are usually very low. By contrast the ingested levels of the precursor, nitrate, and the levels of nitrite and NNC formed endogenously in some human body fluids are much higher. The factors determining the levels of endogenous nitrite and N-nitroso compound formation from dietary nitrate are numerous, some relating to the effects of pre-existing disease states (e.g. achlorhydria, urinary tract infection), some to the metabolic activity of bacteria at certain sites within the body (e.g. the stomach, mouth and urinary bladder) and some to basic mammalian processes (e.g. salivary nitrate recirculation). The complexity of this relationship between nitrate exposure and the potential for disease causation is further increased in many instances by nitrate exposure and metabolism being only one of several suspect aetiological factors. An extreme example is intestinal-type gastric cancer which, although associated in some studies with high nitrate exposure (Hartman 1983), is a disease with a very complex aetiology (Correa 1985). In this, a multitude of *environmental* (e.g. irritants [NaCl, aspirin], nutrient deficits [protein, vitamins, antioxidants]) and *host* (bile reflux, loss of stomach acidity, *Helicobacter* {formerly Campylobacter} *pylori* infection) factors are thought to be important, some of which are quite independent of the potential for nitrate to be a precursor to nitrite and genotoxic NNC (Preussmann and Tricker 1988).

With the complexity of the situation described above, before it is possible to realistically assess the various potential health risks associated with nitrate exposure, it will be very important to have a clear understanding of the pharmacology of nitrate and the factors involved in its metabolism. In 1982, Hartman carefully reviewed much of what was known at the time regarding the pharmacology of nitrate in relation to the causation of human disease. In this chapter we will not describe in detail much of this previously reviewed information, but will concentrate on those areas where newer studies have provided further insights.

7.2 DIETARY EXPOSURE TO NITRATE AND NITRITE

Dietary exposure to nitrate is very variable between individuals, regions and countries. However, average dietary nitrate exposure levels have been quoted for several, ranging in western populations from 0.5 mmol/day/person for Norway to 2.1 mmol/day for the Netherlands (Hartman 1982). Intermediate values have been quoted for the UK and the USA: 0.93–1.9 mmol/day (Hartman 1982, Knight *et al.* 1987, MAFF 1987) and 0.63–1.6 mmol/day (Hartman 1982), respectively. The values for the UK and the USA are based on food consumption data and are probably not the best estimates possible. Nitrate from drinking water contributes a very variable percentage of the total nitrate exposure since water nitrate levels vary considerably between regions. In the USA, 0.7% and also 2.7% of total exposure (US National Academy of Sciences, 1981 from MAFF (1987) have been reported (White 1975); in the UK, values between 0.3% and 23.4% (Knight *et al.* 1987) and between 14 and 25% (MAFF 1987) have been quoted. For UK populations, nitrate from drinking water has been shown to be a poor indicator of total nitrate exposure except for those regions where water nitrate levels exceed 50 mg/l (Caygill *et al.* 1986), representing about 33% of total nitrate exposure.

Various methods of assessing and comparing the nitrate exposure of populations have been used in these studies, including the use of dietary questionnaires and food tables (MAFF 1987, Knight *et al.* 1987), measurement of nitrate and nitrite in saliva samples (Forman *et al.* 1985) and the measurement of nitrate in 24 h urine collections (Chilvers *et al.* 1984, Caygill *et al.* 1986). Dietary questionnaires are notoriously unreliable and time consuming. They are confounded largely by poor dietary recall (Bingham 1987) and the unreliability of food tables since nitrate levels in drinking water and vegetables vary considerably from region to region and with the time of year (White 1983, Knight *et al.* 1987). It was thought that the measurement of salivary nitrate and/or nitrite may overcome some of these problems. However, nitrate and nitrite concentrations in saliva fluctuate markedly throughout the day, and there is little evidence that such determinations are likely to reflect dietary levels reliably (Packer *et al.* 1989). By contrast, 24 h urinary nitrate determinations probably provide the best method of assessing the nitrate exposure of a population since these have been repeatedly shown to represent reliably *ca* 60% of total dietary nitrate exposure (Packer *et al.* 1989). Although somewhat more difficult to collect than spot saliva samples, in our experience this is not a major difficulty. Thus the values reported above for the UK are probably underestimates. Studies of our own, based on urinary excretion of nitrate, have demonstrated that dietary exposure is probably more likely to average 2.2 mmol/day in the UK (Packer 1989).

Dietary exposure to nitrite is low by comparison with exposure to nitrate, with estimates for western populations varying between 0.0024 mmol/day for Norway and 0.097 mmol/day for the Netherlands. Values of between 0.007 mmol/day and 0.054 mmol/day have been quoted for the USA, and values around 0.02 mmol/day (0.007–0.018 (MAFF 1987), 0.03 (Knight *et al.* 1987)) for the UK. Although all of these values have been determined using food consumption data, more reliable figures would be difficult to obtain since nitrite, once ingested, undergoes a series of transformations (see below) and is ordinarily not excreted as nitrite in urine.

7.3 NITRATE AND NITRITE EXPOSURE FROM ENDOGENOUS MAMMALIAN BIOSYNTHESIS

The extent to which endogenous nitrate biosynthesis contributes to overall nitrate exposure in normal healthy humans has been estimated variously to be in the range of 0.9–1.28 mmol/day (Mitchell *et al.* 1916, Green *et al.* 1981, Wagner *et al.* 1983, Leaf *et al.* 1987, Tannenbaum *et al.* 1978a) mostly in US-based studies. Some considerable doubt exists, however, as to the reliability of these estimates. Hartman (1982) argues from several points of view that most of the early studies are likely to have given overestimates, and from the point of view of measurements of the total oxidant load entering the blood stream suggests values more in the range 0.15–0.3 mmol/day.

Many of his arguments are probably still valid in relation to several of the more recent studies (Wagner *et al.* 1983, Leaf *et al.* 1987). Studies of endogenous nitrate biosynthesis have usually either relied upon the estimation of excess 24 h urinary nitrate excretion over ingestion in subjects on low-nitrate diets or/and employed the technique of isotope dilution based on ^{15}N-nitrate administration. For either of these methods of determining endogenous nitrate biosynthesis to have been reliable they would have been greatly dependent upon the accuracy of the nitrate determinations on the low-nitrate food components. For complex mixtures of dietary items such determinations are difficult and prone to underestimation, a point that has already been raised in relation to such studies (Witter *et al.* 1979a; Radomski *et al.* 1978). The reliance on literature values for the nitrate content of some dietary items in studies of this type (Mitchell *et al.* 1916, Green *et al.* 1981, Wagner *et al.* 1983) is even more hazardous since the nitrate content of most food items varies considerably depending on manufacturing practices and growing season. In addition it is arguable that several of these studies may have been further compromised by the body tissues of the test subjects not being in complete nitrate equilibrium. For example, we have recently confirmed earlier reports (Bartholomew and Hill 1984) which demonstrated that a period of at least three days is required for urinary nitrate concentrations to reach a relatively stable basal level (Packer *et al.* 1989) following the initiation of stringent dietary nitrate restriction after free-choice diets. As Hartman points out, in cases where a sudden switch from a high to a low level of nitrate ingestion has taken place there could be a considerable outflow from tissue compartments over a period of several days.

This is a very important point and although the authors would agree with the point of view that a major loss of body nitrate occurs very rapidly (within 48–60 h) following stringent dietary nitrate restriction, it is by no means clear that further pools of less rapidly equilibrating nitrate are not still to be lost over more extended periods of time (Packer *et al.* 1989). We further suggested that since it seems reasonable to assume that the half-lives for both the emptying and the filling of these pools would be of similar magnitude, the relatively short exposures to high nitrate doses or ^{15}N-labelled nitrate that have been used in every study to date (Wang *et al.* 1981, Whelan 1935, Greene and Hiatt 1953) would not have been expected to affect significantly the proportion of ^{15}N to ^{14}N nitrate or the overall nitrate burden in such slowly equilibrating pools. A point that has never been adequately considered in this light, except by Hartman, is that nitrate has actually been shown to be incorporated into the carcasses of dogs and rats exposed to high nitrate levels for relatively short periods of time (16 and 14, days respectively) (Wang *et al.* 1981, Whelan 1935). It has been stated several times or assumed that this incorporation is low and thus there are no slowly equilibrating body pools (Green *et al.* 1981, Tannenbaum 1979, Wagner *et al.* 1983). However, such low incorporation would in fact be expected of such body pools, given that the exposure time is relatively short (Packer *et al.* 1989). Indeed Hartman pointed out, from the data of Whelan (1935), that based on tissue levels of nitrate in the dog, a comparable level of nitrate 'stored' in a 70 kg human adult would total 6.7 mmol. These data are based on animals previously fed with unspecified nitrate intakes (but described as 'not appreciable'), and relate to a standard diet which for a carnivore might be expected to be lower in nitrate than a human diet.

Model calculations of our own would suggest that if there are slowly equilibrating body pools of nitrate, these could take a substantial period of time to be depleted, largely depending on their half-life. The half-life itself may well depend on the degree to which the dietary nitrate levels are reduced during the experiment by comparison with those prior to the initiation of the experiment. The level of nitrate appearing in urine as a consequence of such a process would depend both on the half-life of this nitrate store and the size of the original pool. The size of the original pool may itself depend on the level of regular dietary nitrate intake prior to the experiment. In Fig. 7.1 we have fairly arbitrarily assumed a slowly equilibrating pool

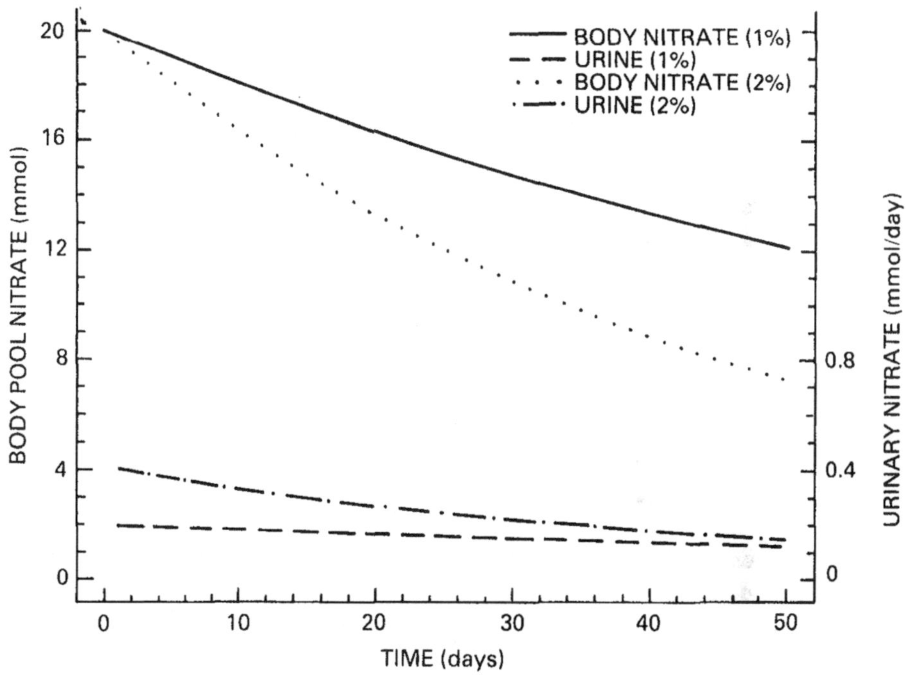

Fig. 7.1 — Mathematical model of the nitrate depletion of a slowly equilibrating body pool and its effect on urinary nitrate.

of 20 mmol, which distributed throughout a body mass of 70 kg would give a tissue concentration of 286 μM. This is only three times the tissue nitrate levels reported by Whelan (1935) for the dog, and these would be expected to underestimate the human levels for the reasons given above. A small daily percentage loss of this pool appearing in urine would give rise to an exponential decay of body nitrate levels and significant urinary levels of nitrate that would persist for many days, 1% loss and 2% loss giving 200 μmol/day reducing to 122 μmol/day and 400 μmol/day reducing to 148 μmol/day respectively over a *50-day* period. It would be impossible to distinguish between the contribution of a process such as this to the urinary nitrate levels of individuals on low-nitrate diets and that of endogenous nitrate biosynthesis from L-arginine, using the types of experimental protocol that have been utilized so far.

More recently, from our own carefully controlled nitrate balance studies, we have calculated a value closer to the range 0.29–0.38 mmol/day (Packer *et al.* 1989) for endogenous nitrate biosynthesis. This estimate more closely agrees with those derived by Hartman (1982) from his review of a wide range of nitrate-balance studies and was derived from regression analysis of a substantial body of data derived from 24 h urinary nitrate excretion and corresponding 24 h nitrate intakes for 13 separate healthy subjects maintained on an extremely rigorously standardized and spartan dietary regime. The result of this regression analysis demonstrated a highly significant positive correlation between the two variates (24 h nitrate intake and 24 h nitrate excretion), with an average of 55% of dietary nitrate being recovered across a range of intakes (0.3–2.1 mmol/day) and an average background excretion of 0.22 mmol apparently independent of immediate dietary nitrate intake, which is the maximum that could be attributed to endogenous synthesis in our study. The value for the recovery of dietary nitrate in urine determined in this study accords well with those obtained in earlier studies utilizing labelled (^{15}N) challenges (e.g. 53% in the study of Green *et al.* (1981)). This lends confidence to the details of the relationship derived between nitrate intake and excretion in our study:

$$(\text{Nitrate}_{\text{excretion}} = 0.55\ \text{Nitrate}_{\text{ingestion}} + 0.22,\ \text{in mmol})$$

Extremely low dietary nitrate intakes were not required for our analysis, avoiding to some degree the problems inherent in switching individuals from high to very low nitrate intakes in relation to the potential for increased leaching of nitrate from less accessible body pools.

All estimates, including our own, of endogenous nitrate biosynthesis have included a correction factor to make allowance for incomplete (60%) recovery in urine. From the results of a very recent study such a correction factor would now seem to be valid (Leaf *et al.* 1990).

We would conclude therefore that mammalian endogenous nitrate biosynthesis is of considerably less significance to urinary nitrate levels and overall nitrate exposure than dietary sources in *normal healthy* individuals, probably being somewhere in the range 0.2–0.4 mmol/day (Packer *et al.* 1989). The extent of endogenous nitrate biosynthesis is an important consideration if 24 h urinary nitrate output is to provide a reliable measure of dietary nitrate exposure in normal, healthy human beings as has been recommended by several workers.

All of this is not to say that endogenous synthesis is not important in humans. In several instances of disease this process may become very important. Stimulation of the immune system has been shown in some cases to effect the normal pharmacology of nitrate in man by increasing urinary nitrate levels, presumably as a consequence of stimulated endogenous nitrate biosynthesis. During the course of nitrate-balance studies in adult humans (Wagner *et al.* 1983) urinary nitrate concentration was discovered to increase ninefold in one subject, who developed a fever and a non-specific intestinal diarrhoea. Similarly, infants suffering from diarrhoea have also been shown to have elevated urinary nitrate levels (Hegesh and Shiloah 1982).

Similarly, giardiasis patients suffering from chronic or acute diarrhoea as a result of infection of the mucosa of the bowel with the protozoa *Giardia lamblia* have also been shown to have elevated levels of nitrate in their urine (Wettig *et al.* 1990). It is interesting to note that all of these cases of increased endogenous synthesis in humans relate to gastrointestinal (GI) infections. It remains to be shown whether other infections also demonstrate a similar effect or whether it is restricted to GI infections. In the case of viral meningitis, Schulz *et al.* (1990) have reported no increase in endogenous nitrate synthesis. This formation of nitrate has been shown in several *in vitro* studies to occur concomitantly with nitrite and NNC formation, both of which could have important toxicological consequences.

7.4 NITRATE – ITS METABOLISM AND TRANSPORT IN BODY FLUIDS

Following ingestion, nitrate passes from the mouth into the stomach and is then rapidly and actively absorbed into the blood from the proximal small intestine. Serum nitrate is then actively secreted into saliva (Spiegelhalder *et al.* 1976), and by more passive mechanisms passes into several other body fluids including, sweat (Bartholomew and Hill 1984), gastric juice (Ruddell *et al.* 1978), gastric mucin (Mueller and Henninger 1985), milk (Hartman 1982) and possibly also vaginal secretions (Alsobrook *et al.* 1974). Ultimately a significant proportion of ingested nitrate (*ca.* 55–65%) is excreted in the urine (Bartholomew and Hill 1984, Green *et al.* 1981, Wagner *et al.* 1983, Packer *et al.* 1989). Typical body-fluid levels are shown in Table 7.1.

7.4.1 The fate of nitrate in the oral cavity and its recirculation in saliva

Nitrate enters the mouth during the first pass of food and water through this cavity, but has a relatively short residence time there. Once in the upper small intestine, dietary nitrate is rapidly absorbed into the blood, and a proportion of it is subsequently recirculated by an active process in saliva into the mouth. In this way salivary recirculation continuously cycles nitrate through the oral cavity, discrete nitrate challenges giving rise to a cycling in salivary nitrate concentration. This process considerably extends the contact time of the oral cavity with dietary nitrate after each meal. Rough estimates of *ca.* 25% have been made for the proportion of a given nitrate dose which undergoes recirculation in this way (Spiegelhalder *et al.* 1976, Bartholomew and Hill 1984). The concentration in saliva peaks approximately 1 h after ingestion (Bartholomew and Hill 1984) and then subsequently decreases in a non-linear fashion with a superimposed cycling in concentration due to salivary recirculation (Walters and Smith *et al.* 1981). Although nitrite is present in whole saliva, parotid duct saliva has been shown to be free of nitrite (Tannenbaum *et al.* 1974, Granli *et al.* 1989). The origin of nitrite in saliva appears to be predominantly the bacterial reduction of nitrate to nitrite by the oral flora.

The microbial flora of the mouth is very complex and in most instances structured into communities with species from a wide range of distinct genera tending to occupy their own particular niche (Theilade 1989). Several of the species and strains of bacteria in these communities will possess nitrate reductase activity (Table 7.2), and,

Table 7.1 — Nitrite and nitrate concentrations in various body fluids

	Nitrite (μM)	Nitrate (μM)
Saliva		
Normal diet[a]	54	NA
Normal diet[b]	0–216	113–629
(following high[c]	2170	9670
nitrate challenges)[d]	2830–11080	2580–7260
25 mg NO_3^- challenge[e]	400	1000
170 mg NO_3^- challenge[e]	1600	NA
Gastric juice, non-fasting		
Normal[e] 25 mg NO_3^- challenge	NA	500
Normal 170 mg NO_3^- challenge	NA	2300
Normal[f] (normal diet)	<1–19	168–175
Achlorhydric (normal diet)	19–111	94–203
Normal[g]		
200 mg NO_3^- challenge	4–198	NA
BI		
200 mg NO_3^- challenge	28–652	NA
BII		
200 mg NO_3^- challenge	83–2030	NA
CAG		
200 mg NO_3^- challenge	9–1871	NA
PGV		
200 mg NO_3^- challenge	4–461	NA
Gastric juice, fasting		
Normal (pH<2.5)[h]	1.7±0.3	NA
Achlorhydric (pH>5)	30.4+4.9	NA
Normal (pH<4)[i]	2(0–39)	108(27–274)
Achlorhydric (pH>5)	26(2–102)	169(18–387)
Normal (pH<4)[j]	6.5(0–24)	NA
Achlorhydric (pH>5)[j]	22(0–52)	NA
pH variable	8.7(0.5–48)	NA
Normal (pH<5)[k]	7±7	520±320
Achlorhydric (pH>5)	613±864	520±460
Normal[l]	2.6	NA
PA (ph>5)	12	NA
Normal (ph<4)[m]	ND	114–486
PA (pH>6)	2–129	0–122
HGGA (pH>7)	72–309	0–386
Normal[g]	ND–106	NA
BI	2–74	NA
BII	17–398	NA
CAG	17–96	NA
PGV	ND–196	NA
Ileostomy fluid[n]	9.5(<5–15)	<5(<507)
Faeces[n]	12(5–19)	5(<4–14)
Urine		
Neoplastic bladder		
Disease[o]	ND–435	NA
Hemi- and paraplegia	ND–435	
Sterile[p]	1±5	1080±740
Bacteriuria	46±79	1030±1580
Egyptian controls[q]	only 2/27 with NO_2^-	1840±1075
Bilharzia patients	72±126	1469±1397
Serum[r]		40

[a] Tannenbaum *et al.* (1978b), [b] Tannenbaum *et al.* (1974), [c] Ishiwata *et al.* (1975), [d] Tannenbaum *et al.* (1976), [e] Hill (1979), [f] Caygill *et al.* (1987), [g] Adam *et al.* (1989), [h] Ruddell *et al.* (1976), [i] Mueller *et al.* (1986), [j] Mueller *et al.* (1984), [k] Tannenbaum *et al.* (1979), [l] Ruddell *et al.* (1978), [m] Dolby *et al.* (1984), [n] Saul *et al.* (1981), [o] Hicks *et al.* (1977), [p] Ohshima *et al.* (1987), [q] Tricker *et al.* (1989), [r] Hartman (1982).

PA=pernicious anaemia, HGGA=hypogammaglobulinaemia, BI=Billroth I, BII=Billroth II, PGV=proximal-gastric vagotomy, CAG=chronic atrophic gastritis, ND=not detected, NA=no analysis.

Table 7.2 — Members of the bacterial flora of the mouth. No. of species is given in parenthesis

Aerobes, facultative anaerobes, microaerophils		Anaerobes	
Streptococcus	(11)	*Peptostreptococcus*	(2)
Stomatococcus	(1)	*Actinomyces*†	(2)
Actinomyces†	(3)	*Propionibacterium*†	(1)
Arachnia†	(1)	*Bifidobacterium*	(1)
Corynebacterium†	(1)	*Eubacterium*†	(5)
Rothia†	(1)	*Veillonella*†	(3)
Lactobacillus†	(7)	*Bacteroides*†	(19)
Neisseria†	(4)	*Mitsuokella*	(1)
Branhamella†	(1)	*Fusobacterium*†	(3)
Haemophilus†	(5)	*Leptotrichia*	(1)
Actinobacillus†	(1)	*Wolinella*	(2)
Eikenella	(1)	*Selenomonas*†	(1)
Capnocytophaga	(3)	*Centipeda*	(1)
Campylobacter†	(2)		
Simonsiella†	(1)		

Adapted from Theilade (1989), where full species list is cited.
† Genera with nitrate-reducing species (several sources).

depending on such factors as their level of carriage within the flora, their spatial distribution in relation to salivary nitrate flow, and their specific nitrate reductase activities in relation to those environmental factors modulating expression, lead to a particular level of conversion of nitrate to nitrite in the saliva of an individual. It has further been shown that salivary flow, in turn, may be influenced by climatic conditions. With higher external temperatures it has been suggested that salivary flow might be reduced, resulting in a more concentrated saliva (Eisenbrand *et al.* 1980), which would both increase the activity of bacterial enzymes and substrate concentrations, resulting in increased nitrite concentrations (Boss *et al.* 1988).

If the species distribution of the oral flora and its metabolic activity is sensitive to external influences, and in particular to dietary composition, then these, in turn, may become important factors determining nitrate to nitrite conversion in the mouth. It has been shown, for example, in the field of dental caries' research, where the effects on the flora of such influences have been more carefully examined, that higher dietary sucrose levels are likely to select, from amongst the oral flora, for those organisms which tend to be acid producing (and cariogenic) and acid tolerant (McDermid *et al.* 1986). By analogy, the carbohydrate composition and nitrate content of the regular diet may also influence the species distribution and metabolic activity of the flora in relation to nitrite formation. The effect of such factors is considered in greater detail in the review by Hartman (1982).

Another factor which requires consideration is the effect of bacterial *nitrite* reductase on salivary nitrite levels. The effect of this enzyme activity will be to reduce the availability of the substrate, nitrite, for subsequent N-nitrosation reactions and genotoxic NNC formation. Several of the organisms with nitrate reductase activity will also possess nitrite reductase activity and it will be the balance of the cumulative activities of these two classes of enzyme within the members of any one oral flora that will tend to determine overall salivary nitrite levels. The relative expression of these two classes of enzyme will be under environmental control, with such factors as the availability of oxygen and suitable electron donors playing a major role (Hochstein and Tomlinson 1988, Knowles 1982, Payne 1973, Stewart 1988). These factors have been given little consideration with particular regard to the oral cavity, but have been given closer consideration with respect to the achlorhydric stomach (see below). None of these factors has been systematically investigated with respect to the precise details of the species composition and metabolic activities of an individual's flora, in relation to their propensity for converting nitrate into nitrite in the mouth. However, it may be inferred that considerable differences exist between individuals in this respect since large inter-individual differences in salivary nitrate to nitrite conversion are possible (Table 7.3).

Table 7.3 — Average conversion of salivary NO_3^- to NO_2^- for eleven volunteers in the 24 h following consumption of a fixed low-nitrate diet and following three different nitrate challenges (*ca.* 1.0 mmol). From Packer *et al.* (1989)

Subject	No dose	Water	Lettuce subject	Ham	Overall summary
1	$46(29)^{8}$	$21(10)^{10}$	$15(18)^{10}$	$67(63)^{10}$	$28(33)^{41}$
2	$83(46)^{7}$	$53(36)^{7}$	$86(46)^{8}$	$44(75)^{5}$	$72(54)^{27}$
3	$48(45)^{4}$	$61(34)^{4}$	$100(0)^{4}$	$100(20)^{8}$	$87(46)^{20}$
4	$61(25)^{6}$	$43(31)^{8}$	$33(87)^{7}$	$22(50)^{7}$	$44(38)^{28}$
5	$43(37)^{6}$	$36(11)^{8}$	$36(18)^{5}$	$29(38)^{7}$	$38(17)^{26}$
6	$79(25)^{7}$	$52(14)^{6}$	$55(09)^{7}$	$72(29)^{6}$	$61(23)^{27}$
7	$64(42)^{7}$	$51(26)^{6}$	$55(42)^{7}$	$43(14)^{6}$	$53(31)^{26}$
8	$67(47)^{8}$	$38(19)^{5}$	$55(45)^{7}$	$40(12)^{8}$	$46(34)^{20}$
9	$72(53)^{6}$	$82(03)^{6}$	$85(35)^{6}$	$87(13)^{5}$	$83(32)^{22}$
10	$24(11)^{9}$	$29(05)^{6}$	$19(13)^{7}$	$9(04)^{8}$	$22(15)^{30}$
11	$48(38)^{7}$	$25(19)^{6}$	$35(46)^{6}$	$50(52)^{5}$	$42(38)^{25}$
Overall dose summary	55(40)	38(28)	47(57)	46(57)	46(47)

Key: $a(b)^c$; a=median, b=inter-quartile range, c=number of observations.
Conversion=$NO_2^-/(NO_3^-+NO_2^-)$.

Further, despite considerable temporal differences in salivary nitrate and nitrite concentration within individuals, several studies, including our own, have demonstrated that subjects tend to be consistently high, medium or low nitrate converters (Packer *et al.* 1989, Boss *et al.* 1988) and this has been attributed to differences in the metabolism of the oral flora. Similar distinctions between individual monkeys have been made in animal studies (Hayashi *et al.* 1978). Speigelhalder *et al.* (1976) estimated that in humans, 20% of the 25% of dietary nitrate undergoing salivary recirculation was converted to nitrite. This represents 5 mol% of overall nitrate intake. Similar estimates were obtained by Stephany and Schuller (1980), who reviewed data from the Netherlands, Japan and FRG. They calculated that salivary nitrate to nitrite conversion represented on average 6.3 mol% of dietary nitrate. In our own laboratory studies we have found a somewhat higher average of 11.5 mol% for a UK population. An important point to reiterate here, however, is that inter-individual differences in this parameter are very large: the two extremes for the median percentage of nitrite in saliva of 22% and 87% (Table 7.3) give a range in our study of 5.5 to 22 mol% of total dietary nitrate. Such differences in the ability of the oral flora of different individuals to reduce nitrate to nitrite may have important epidemiological consequences since nitrite, not nitrate, is the ion of toxicological significance.

7.4.2 Fate of nitrate in the stomach

Nitrate reaches the stomach as a consequence of at least two and possibly three distinct processes. That is: (1) direct ingestion of nitrate-containing dietary items; (2) via the recirculation of nitrate in swallowed saliva; and (3) possibly as a result of nitrate from serum being pumped actively into gastric secretions. In contrast to the mouth, nitrate entering the stomach with dietary items will have a much longer residence time since significant absorption of nitrate does not occur from the stomach (Witter *et al.* 1979b), and the half-lives of the solid and liquid components of a meal leaving the stomach for the duodenum are about 40 and 80 minutes respectively (Malagelada *et al.* 1984, Mayer *et al.* 1984). The nitrate swallowed with saliva, which is also usually accompanied by nitrite (see above), is another major source of these ions in the stomach and indeed is the primary source of nitrite in the normal acid stomach. As in the mouth, this source extends in time the exposure of the stomach to nitrate and nitrite from discrete nitrate doses. Hartman (1982) in his review considers that nitrate entering the stomach with *gastric* secretions could provide a further contribution to gastric juice nitrate which in some instances exceeds that from saliva. His argument is by analogy with iodide transport, and nitrate transport in rats. Definitive assessments of the relative importance of this source of gastric nitrate, however, are lacking. The recent studies of Mueller *et al.* (1985, 1986), however, has shown that the nitrate concentrations in gastric mucin may exceed those in corresponding gastric juice samples, probably representing the outcome of an active secretory process for nitrate.

By contrast with nitrate, nitrite has a very short half-life in the normal stomach (<10 min.). Using the dog stomach as a model (Licht *et al.* 1986), three processes have been shown to contribute to this. These are predominantly gastric absorption, with minor contributions from the dilution of stomach contents and the chemical reactions of nitrite. Pentagastrin stimulation of gastric acid secretion was found to double the half-life of intragastric nitrite. This was attributed either to a decrease in the permeability of the stomach to nitrite or to an active secretion of nitrite.

As stated previously, in the normal acid stomach (pH range 1–3), intragastric nitrite is largely of salivary origin. However, in those individuals with impaired gastric acid secretion (hypochlorhydria and achlorhydria) the intragastric interconversion of nitrate to nitrite by bacteria provides a further and probably dominant source of nitrite. The normal stomach is acid and therefore essentially sterile, with counts of recoverable luminal bacteria normally less than 10^3/ml of gastric contents. These bacteria almost certainly represent a transient flora with little or no metabolic activity at these pHs. Gastric acid secretion, however, can become impaired by any one of a range of phenomena, including gastric surgery (partial gastrectomy, vagotomy), treatment with H_2 receptor antagonists (cimetidine, ranitidine), metabolic disorders (pernicious anaemia, hypogammaglobulinaemia), chronic poor nutrition and aging. Once gastric acid production is sufficiently impaired and the resting gastric pH is consistently above 4, a resident microbial flora develops, with counts of up to 10^8–10^9 bacteria/ml being reported (Drasar *et al.* 1964, Leach and Hill 1990). A considerable range of genera and species are represented amongst the bacteria which have been isolated from such achlorhydric and hypochlorhydric gastric juices, and nitrate reductase activity is a feature of several of the species and strains which have been isolated (Table 7.4). Overall, the counts of nitrate-reducing bacteria increase as the intragastric pH rises, in parallel with the total counts of bacteria.

With the nitrate entering the stomach both as a dietary constituent and as a component of salivary recirculation, the action of this nitrate-reducing component of the flora has the effect of substantially increasing the intragastric concentration of nitrite above that which ordinarily results from the swallowing of saliva (Adam *et al.* 1989; see above). A clear relationship has therefore been broadly demonstrated between gastric pH, counts of nitrate-reducing bacteria and the ensuing intragastric nitrite levels. These last-mentioned may become increased from the levels of around 3 μM found in normal (acid) gastric juice, by a factor of 10 or more, to concentrations in the range 30–300 μM in the gastric juice of achlorhydrics (Table 7.1). However, considerable intra- and inter-individual variation in intragastric nitrite concentration has been demonstrated, particularly amongst achlorhydric subjects. Coupled with such non-microbial factors as the rate of gastric emptying and the dietary nitrate load, the species composition and metabolic activity of the individual's gastric flora will play a dominant role in determining their intragastric nitrite concentration. This will in turn be determined by environmental factors (particularly dietary ones), many of which have already been discussed above in relation to nitrite formation in the oral cavity. However, in the case of the stomach, an extra factor of some considerable significance may be the precise gastric pH and the degree to which it varies during the course of the day.

The broad dependence of intragastric nitrite concentration on gastric pH, even within the range 4–8, demonstrates the importance of this factor in modulating the composition and/or metabolic activity of the gastric flora (Meyrick-Thomas *et al.* 1987, Milton-Thompson *et al.* 1982, Mueller *et al.* 1983, Mueller *et al.* 1984).

Table 7.4 — Members of the bacterial flora of the achlorhydric stomach

Aerobes, facultative anaerobes, microaerophils	Anaerobes
Streptococcus spp†	*Peptostreptococcus* spp
Enterococci†	*Clostridium* spp†
Staphylococcus spp†	*Peptococcus* spp†
Micrococci†	*Bifidobacterium* spp
Corynebacterium spp†	*Veillonella* spp†
Diphtheroides	*Bacteroides* spp
Lactobacillus spp†	*Fusobacterium* spp†
Neisseria spp†	*Lactobacillus* spp†
Escherichia coli†	*Sarcina* spp
Haemophilus spp†	
Citrobacter spp†	
Klebsiella spp	
Flavobacterium spp†	
Helicobacter pylori	
Bacillus spp†	
Aeromonas spp†	
Pseudomonas spp†	
Enterobacter spp	
Proteus spp†	
Hafnia spp†	

Adapted from Leach and Hill (1990).
† Genera with nitrate-reducing species (several sources).

Again, as with the oral cavity, none of these factors has been systematically investigated in relation to intragastric nitrite formation. However, it is clear that differences between the species composition of different gastric floras will have a profound effect on individual levels of intragastric nitrite accumulation. Further, the availability of appropriate electron donors within the gastric milieu for the different bacterial nitrate and nitrite reductases will be likely to significantly affect the balance of nitrate reduction in favour, or otherwise, of intragastric nitrite accumulation (Mueller *et al.* 1983, 1984, Forsythe and Cole 1987, Forsythe *et al.* 1988).

The metabolism of nitrate by bacteria in the achlorhydric stomach has recently been demonstrated to result in very significant losses of dietary nitrate when the recovery of discrete challenges was measured in urine. Ordinarily about 55% of a discrete nitrate challenge is recovered on average in urine. Achlorhydric pernicious anaemia patients were shown to give much poorer urinary recoveries of about 33%. This represents a major loss of nitrate (23%) to bacterial metabolism in the stomach in this extreme group of achlorhydric individuals (Packer *et al.* 1990). It would be expected that a similar effect would be seen in other achlorhydric and hypochlorhydric groups, the levels of loss differing only in degree dependent upon the precise levels of intragastric bacterial nitrate reductase activity which begins to increase above pH 4 (Milton-Thompson *et al.* 1982). It is interesting to consider these results in the light of the most recent study on the half-life of nitrite (Licht *et al.* 1986) in the normal dog stomach. Licht *et al.* concluded from their study that the major factor responsible for the short half-life of nitrite in the normal stomach was gastric absorption, with only a minor contribution from the reactions of nitrite. This implies that under these circumstances any nitrite formed in the normal acid stomach would be promptly absorbed into the blood, undergo oxidation back to nitrate by reaction with oxyhaemoglobin, and not result in losses to urinary recovery of dietary nitrate challenges. By contrast, in the achlorhydric stomach it is now clear that any nitrite formed there, or swallowed with saliva, will have a much greater propensity for further reaction than in the normal stomach, this now becoming the dominant process of nitrite removal. The reactions to which this nitrite would be prone would include further bacterial reduction to ammonia and/or gaseous nitrogen oxides, and bacterially catalysed N-nitrosation. This latter may relate itself quite closely to the bacterial formation of nitrogen oxides from nitrite (see section 7.5).

7.4.3 The fate of nitrate in the intestinal tract

On leaving the stomach, nitrate is thought to be actively and reasonably completely absorbed in the proximal small intestine, there having been little absorption from the stomach itself (Witter *et al.* 1979a, 1979b, Witter and Balish 1979, Hartman 1982). In their pharmacokinetic model of nitrate absorption and metabolism, Schultz *et al.* (1985), however, assume a passive absorption of nitrate from the proximal small intestine. Once absorbed, nitrate enters the blood and becomes available for secretion into various body fluids, and ultimately for excretion in urine. From the point of view of processes affecting nitrate pharmacology it is important now to consider the *large* intestine as a site of nitrate metabolism and loss. It is likely that losses do occur here as a consequence of the metabolic activity of its abundant bacterial flora. It is unlikely, however, that nitrate enters this organ as a consequence of the bulk flow of contents down the intestine since the absorption from the proximal small intestine is rapid. Studies in rats suggest that nitrate diffuses back from blood into the large intestine (Witter *et al.* 1979b, Ishiwata and Tanimura 1982, Thayer *et al.* 1982). The role of *bacterial* metabolism in the large intestine in contributing to the loss of nitrate is suggested by a comparison of the urinary recoveries of discrete nitrate challenges between conventional flora and germ-free rats. Bacterial metabolism in the large intestine was shown to account for probably about half of the 40–45% of the total loss of nitrate from the body prior to excretion in urine (Schultz *et al.* 1985). This compares quantitatively with the losses previously described which occur in the achlorhydric bacterially colonized stomach.

7.4.4 The excretion of nitrate in urine

A major clearance of nitrate from the blood occurs in the kidneys. Renal clearance of discrete nitrate challenges has been estimated variously to be in the region of 50% to approximately 90% of an oral dose in the 24 h following ingestion. Hartman concluded in his review (1982) that urinary recovery of nitrate approximates the amount ingested, particularly at high nitrate doses where the recovery may approach 90%. He also suggested that with lower nitrate challenges, nitrate synthesized endogenously as a consequence of mammalian processes may obscure the losses of nitrate to some degree, thus increasing the apparent recovery of challenges. More recent studies using ^{15}N-labelled nitrate have demonstrated both in man and in the rat that 24 h urinary recovery is closer to 50–60% (Leaf *et al.* 1987, Green *et al.* 1981, Wagner *et al.* 1983). In rats the recovery of the labelled nitrate was not significantly different when introduced intravenously or intraperitoneally as compared to orally (Leaf *et al.* 1990).

We have recently demonstrated in very carefully controlled nitrate-balance studies of our own that in man a higher apparent recovery of nitrate challenges (70–100%) will be observed if the background nitrate excretion which is independent of recent nitrate exposure (e.g. endogenous synthesis) is not taken into account. Regression analysis of our data demonstrated as expected that 24 h urinary nitrate excretion was closely correlated with challenge dose across a range of intakes (0.3–2.1 mmol/day) but that there was a background daily excretion of nitrate of 0.22 mmol/day which was not related to recent nitrate exposure, giving an actual recovery of 55% of daily nitrate intake over the entire range of intakes. In later experiments we demonstrated very similar actual recoveries for even higher nitrate doses, 2.5 mmol and 5.0 mmol challenges giving 58% and 51% recovery in urine respectively (Packer *et al.*, 1988).

Renal clearance of serum nitrate has been shown to be a fairly rapid process. Urinary nitrate concentrations apparently reach a maximum approximately 3 to 5 h following the ingestion of discrete nitrate challenges (Bartholomew and Hill 1984), 2.7 mmol and 0.8 mmol challenges resulting in peak urinary nitrate concentrations of about 2.26 mM and 1.13 mM respectively. Despite this, there are likely to be considerable fluctuations in urinary nitrate concentration that are not entirely predictable purely in relation to the magnitude and timing of normal nitrate ingestion. For example, we have demonstrated that urinary nitrate concentration fluctuates with a circadian periodicity in fasting subjects (Packer *et al.* 1989). Such circadian patterns probably continuously underlie urinary nitrate excretion but are masked to a large degree by the fresh input of nitrate with each meal. Further, we demonstrated that with individuals on carefully controlled and relatively low dietary nitrate (*ca.* 0.6) intakes supplied solely by a formula-based diet, urinary nitrate concentrations fluctuated dramatically over 13 days' with little obvious dependence on the timing of 'meals'. Increasing nitrate intake by a further 1 mmol at midday on three separate days during the experiment did not noticeably affect the apparent lack of pattern in urinary nitrate excretion. It is probable that the time base of the urine collections will be a determining factor in the likelihood of observing patterns in urinary excretion that relate to the timing of dietary intakes. In our study only six urine collections were made every 24 h on average, and in previous studies a much shorter period between collections had been used. In the latter case, the effect on urinary nitrate concentration of discrete challenges would be more noticeable. Only when our data were expressed as nitrate excreted over time periods of 12 h or more was it possible to observe the effects of the 1 mmol challenges as an increase in urinary nitrate.

Urine ordinarily contains little or no nitrite. However, with diseases that result in the colonization of the bladder by bacteria, nitrite is quite frequently found. Hartman (1982) suggested from his review of the literature that urinary tract infections are not uncommon, and in many cases may be asymptomatic, with nitrite in infected urines ranging from 0.22 to 1.7 mmol/day. In a more recent study of patients with urinary tract infections, urinary nitrite concentrations ranged from not detectable to 330 μM (Ohshima *et al.* 1987). In this study, nitrite in infected urine appeared to depend heavily on the species of microorganism colonizing the bladder, *Escherichia coli* being both the organism predominantly isolated from these urines and the organism associated with urinary nitrite formation. Nitrite was also found in urines infected with *Klebsiella* spp, *Proteus* spp and *Staphylococcus* spp. Urinary tract infections can be a particular and persistent problem for certain groups of individuals. Thus Hicks *et al.* (1977), in their study of hemi- and paraplegics, reported nitrite concentrations in infected urines ranging from not detectable to 435 μM, and similar levels in the infected urines of patients with neoplastic bladder disease. The organisms isolated from urine in these studies were faecal streptococci, coliforms, staphylococci and *E. coli*, with some urines having particularly heavy and mixed growths. Tricker *et al.* (1989), in their study of bilharzia patients, who regularly develop heavy and mixed bacterial infections in their bladders, showed urinary nitrite concentrations to range from not detectable to 978 μmol/day.

7.4.5 Sources and modulation of endogenous nitrate biosynthesis

As discussed previously in relation to whole-organism studies, it has been known for some time that it is likely that there are purely mammalian metabolic processes which are responsible for the endogenous synthesis of nitrate. While the question of the extent of this process in normal healthy humans and rats may not yet be resolved (see section 7.2), it is clear from both whole-animal studies and also from *in vitro* cell culture and whole-tissue experiments that this process is a fundamental part of the metabolism of a range of cell types, and may be modulated by a range of effectors.

It was first suggested that endogenous nitrate biosynthesis may occur as a consequence of bacterial nitrification taking place in the gastrointestinal tract (Tannenbaum *et al.* 1978a). It has subsequently been realized, however, that such a process is unlikely to take place under the anaerobic conditions which occur in much of the GI tract (Witter *et al.* 1979a). Dull and Hotchkiss (1984) and Saul and Archer (1984) in later studies demonstrated the oxidation of ammonia and hydroxylamine to nitrate in *in vitro* incubations of rat liver. This was attributed to the action of both superoxide and hydroxyl free radicals reacting with ammonia to form hydroxylamine and ultimately nitrate. A xanthine–xanthine oxidase reaction was suggested as a possible source of oxidants in these reactions.

More recent studies of endogenous nitrate biosynthesis have concentrated on the process mediated by macrophages which, upon immunostimulation by either the exogenous stimulant *Escherichia coli* lipopolysaccharide (Stuehr and Marletta 1985) or the endogenous lymphokine interferon-γ (IFN) (Stuehr and Marletta 1987), clearly demonstrate nitrate and nitrite formation in the culture medium. Not only is this phenomenon observable with cultured murine macrophages following stimulation but also in whole mice injected intraperitoneally with *E. coli* LPS (Stuehr and Marletta 1985). In the case of the whole-animal experiments, only increases in *nitrate* in the urine and blood were observed, any nitrite formed by the macrophages presumably becoming oxidized by oxyhaemoglobin to nitrate.

Further studies with macrophages in culture revealed that the precursor to nitrate and nitrite was the amino acid L-arginine (Iyengar *et al.* 1987, Hibbs *et al.* 1987) and that the other major product of the reaction was probably L-citrulline. Studies with ^{15}N- and ^{14}C-labelled L-arginine demonstrated that the N atoms of both nitrate and nitrite were derived exclusively from one of L-arginine's two equivalent guanido nitrogens, and that citrulline was the other major product (Iyengar *et al.* 1987, Leaf *et al.* 1990, Marletta 1988). An enzyme activity from macrophages isolated in the 100000 g supernatant has now been demonstrated to produce *nitric oxide* from L-arginine. This enzyme activity was only recoverable from cells which had been previously stimulated with LPS and γ-IFN, being completely absent from unstimulated cells (Marletta *et al.* 1988). Nitric oxide would be expected to be rapidly oxidized by molecular oxygen and then undergo further reaction to form nitrite and nitrate. Nitric oxide is thus a likely intermediate in the formation of these ions by macrophages from L-arginine (Marletta, 1989). The concomitant nitrosation of secondary amines when added to cultures of activated macrophages synthesizing nitrate and nitrite (Miwa *et al.* 1987) has also been attributed to the formation of this intermediate, nitric oxide, (Marletta *et al.* 1988) which on oxidation by molecular oxygen would be expected, amongst other things, to give rise to the active nitrosating agents N_2O_3 and N_2O_4.

The details of the mechanism by which L-arginine is oxidized to citulline and nitric oxide are not clear. Marletta *et al.* (1988), however, have described a detailed mechanistic proposal in which an NADPH-dependent N-hydroxylation reaction results in the formation of N^G-hydroxy-L-arginine. This is then proposed to undergo a series of further reactions which lead first to the elimination of NO and then to the formation of citrulline.

Cell types other than macrophages have also recently been shown to produce NO from L-arginine by a very similar process to that occurring in macrophages, including endothelial cells (Palmer *et al.* 1987, Moncada *et al.* 1988), neural tissue (Knowles *et al.* 1989), neutrophils (McCall *et al.* 1989), and possibly hepatocytes and Kupffer cells (Billiar *et al.* 1989). It has been suggested that the NO produced by each of these cell types serves an intercellular signalling function in neural tissue by stimulating soluble guanylate cyclase activity. Nitrate derived by oxidation of the NO from all of these cell types may add to the total body burden of nitrate (Leaf *et al.* 1990) over and above that which is received with the diet (Leaf *et al.* 1989), but as stated earlier this may only become of particular importance following significant stimulation of certain of these cell types. Indeed even in a study designed to stimulate endogenous nitrate biosynthesis by the vasodilation brought on by a prolonged period of exercise (6 h) only relatively small increases in urinary nitrate excretion were observed (400 μmol) in the 12 h immediately following the initiation of exercise. So far, only the presumed immune stimulation associated with diseases resulting in severe or chronic diarrhoea have been shown to significantly affect urinary nitrate output (see section 7.2). In health it still remains likely from our experience that dietary nitrate exposure is the major contributor to the overall body burden of nitrate in man (see section 7.2 and Packer *et al.* (1989)).

7.5 N-NITROSATION REACTIONS AND THE FATE OF NITRITE IN THE BODY

While exposure to nitrite itself either by ingestion or by endogenous formation (both by bacterially mediated and mammalian processes) within the body has important health consequences in relation to methaemoglobin formation, its most controversial health risk, cancer, relates to its role as a precursor to genotoxic N-nitroso compounds. The formation of N-nitroso compounds is thus an important aspect of human nitrate pharmacology and metabolism.

The process by which N-nitroso compounds are formed, N-nitrosation, can be defined as the replacement of a hydrogen, attached to a nitrogen, by a nitroso group, with the donor of the nitroso group being termed a nitrosating agent. A considerable range of mechanisms of N-nitroso compound formation are known to occur, and these have been the subject of several reviews (Douglass *et al.* 1978, Challis and Challis 1982, Shuker 1988). Not all of the potential mechanisms of nitrosation, however, are likely to be of relevance to NNC formation within the body since the physiology of, and substrate availability within, the various sites where nitrite may be formed (mouth, stomach, urinary bladder, colon, vagina) will be unfavourable for many of them (for a recent review, see Leach (1988)). For these sites with largely aqueous environments the two mechanisms of greatest relevance will be the acid catalysed and the bacterially mediated nitrosation reactions of nitrite, the former being particularly relevant to the normal acid stomach and the latter to most of the other sites.

The acid-catalysed reactions of nitrite are well understood. Nitrite, itself, is not a nitrosating agent, the active nitrosating agents, N_2O_3, N_2O_4, H_2ONO^+ and NO^+, all being derived chemically as a consequence of the behaviour of nitrite in acid solution, the balance between the different chemical species being formed depending on the precise pH. A range of catalysts (e.g. thiocyanate, chloride and some phenolics) and inhibitors (e.g. ammonia, urea and vitamin C) of these reactions are known, several of which may be relevant to modulation of N-nitrosation in the acid stomach (Shuker 1988, Archer 1984).

Bacterially mediated N-nitrosation reactions (Calmels *et al.* 1985, Leach *et al.* 1985, 1987a, 1987b, Leach 1988) having pH optima within the range 6-8 are those most likely to be responsible for NNC formation at sites of neutral pH within the body (for example in the mouth, the achlorhydric stomach and the urinary bladder). Under the physiological conditions which prevail at these sites the acid-catalysed reactions of nitrite are likely to be of little significance. Since these sites are also likely to be colonized by nitrate-reducing bacteria (see above) which will provide elevated levels of the substrate, nitrite, they are of particular interest with regard to N-nitroso compound formation.

Bacterially catalysed N-nitrosation appears to be a secondary feature of the enzymes involved in bacterial nitrate and nitrite reduction, and is therefore restricted, in the first instance, to those organisms possessing these enzymes. However, not every bacterial species normally possessing these reductase enzymes demonstrates significant N-nitrosation activity (Leach and Hill 1990). Unlike the acid-catalysed N-nitrosation reactions taking place in the normal acid stomach which depend largely on relatively well characterized factors such as the nitrite concentration, the pH and the presence of catalysts and inhibitors, the extent of the bacterially mediated reactions will depend to a large degree on the precise composition of the bacterial flora in relation to the N-nitrosation activity of its members. This is likely to be a feature which varies considerably between different individuals, given the great variability in the composition of different gastric floras (Mueller *et al.* 1983, 1984).

At least two distinct mechanisms appear to be employed by different bacterial species in the expression of N-nitrosation activity, one involving nitrate reductase in nitrate respirers such as *Escherichia coli* and a second involving nitrite reductase in denitrifiers such as *Pseudomonas aeruginosa* (Calmels *et al.* 1988, Leach *et al.* 1987b, Ralt *et al.* 1988). Both mechanisms have been shown to display reaction kinetics which are consistent with bacterial enzyme mediation (Calmels *et al.* 1985, Leach *et al.* 1985, 1987a), and this is supported by the fact that the reaction in both cases is dependent on the presence of whole live cells. In the main, the specific N-nitrosation activities displayed by these two groups of organisms are very different (Leach *et al.* 1987a, 1987b). In standardized *in vitro* tests, denitrifiers tend to catalyse much more rapid specific rates of reaction than non-denitrifiers, typically by a factor between 10 and 100. The major reason for the difference in the magnitude of the N-nitrosation rate between these two groups of organisms probably relates to the differential in their abilities to produce nitric oxide from nitrite. As discussed previously, nitric oxide (NO) itself is not a nitrosating agent. However, it readily oxidizes in the presence of traces of oxygen to N_2O_3 and N_2O_4, both of which are potent nitrosating agents at neutral pH in aqueous solution. Although these more oxidized nitrogen oxides react readily with water, producing NO_2^- and NO_3^-, amines with a considerable range of basicity compete efficiently for the nitrosating agents and produce the corresponding NNC (Shuker 1988).

It has become increasingly clear that some non-denitrifying organisms such as *E. coli* produce small amounts of NO from nitrite during the course of nitrate respiration (Goretski and Hollocher, 1988, Ji and Hollocher 1988a). It has been suggested that this occurs as a result of the adventitious activity of *nitrate* reductase acting on *nitrite*. Certainly, the enzyme activity most likely to be responsible for nitrosation catalysis in these organisms appears to be nitrate reductase (Calmels *et al.* 1988, Ralt *et al.* 1988). Furthermore, the anaerobic incubation of these organisms with nitrite has been shown to result in the production of a gaseous species which is itself not a nitrosating agent, but only becomes so once traces of oxygen have been introduced into the incubation mixture. Further experiments have also shown that the nitrosation activity of these bacteria is closely mirrored by their ability to generate NO from nitrite (Ji and Hollocher 1988b). It seems most likely therefore that NNC formation by these organisms involves three distinct steps. The first is the adventitious enzymatic reduction of nitrite by nitrate reductase to produce NO. The second is the purely chemical oxidation of NO by traces of dissolved molecular oxygen to N_2O_3 and N_2O_4. The third is the transfer of the nitroso group from these nitrosating agents to suitable amino substrates.

The N-nitrosation reactions of denitrifiers, on the other hand, appear to be more closely associated with their dissimilatory nitrite reductase, not their nitrate reductase (Calmels *et al.* 1988, Garber and Hollocher 1982). This must certainly be true of those denitrifiers which catalyse N-nitrosation but do not possess a nitrate reductase, for example some *Neisseria* spp (Leach *et al.* 1987a). The normal pathway for the reduction of nitrite by denitrifiers differs considerably from that of non-denitrifiers (nitrate respirers) in that the products in this case are the gaseous oxides of nitrogen, and ultimately molecular nitrogen itself, rather than ammonia. While nitrous oxide (N_2O) is known to be a free intermediate in the reduction of nitrite by denitrifiers, it is not a nitrosating agent. However, some denitrifiers may produce free NO as an earlier product of nitrite reduction. In those denitrifiers in which this is the case, this free NO would then go on to produce the nitrosating agents N_2O_3 and N_2O_4 in the same way as the NO generated by the nitrate respirers; that is, as a consequence of chemical oxidation by traces of molecular oxygen.

It is not clear, however, that all denitrifiers produce free NO. For example, those studies of denitrification which have employed heavy isotope labelled (^{18}O, ^{15}N) substrates have indicated that, even amongst denitrifying bacteria, species differ in the precise details of their denitrification sequence, some liberating free NO and others not (Garber and Hollocher 1981, 1982). In those organisms apparently not producing *free* NO (e.g. *P. aeruginosa*) it is thought that the NO remains bound to the nitrite reductase cytochrome as a ferrous nitrosyl. Such a ferrous nitrosyl would be expected to be itself a good nitrosating agent, by analogy with the well known reaction involving nitroprusside (Averill and Tiedje 1982). Preliminary experiments of our own, however, indicate that even the nitrosating agent elaborated by *P. aeruginosa* behaves in a very similar fashion to mixed oxides of nitrogen (N_2O_3 and N_2O_4) in relation to its relative reactivity towards a number of secondary amines of differing pK_a (Leach and Hill 1990).

Thus it is not yet clear what are the dominant N-nitrosating agents elaborated by different denitrifying bacteria from nitrite. Some insight into this may be provided by the denitrification scheme as proposed by Averill and Tiedje (1982), which helps to explain many of the apparently conflicting results obtained in different studies. In this scheme a range of free and enzyme-bound intermediates linking the reduction of nitrite to nitrous oxide (N_2O) are proposed, alongside a system of appropriate electron transfers. An important aspect of this scheme is that certain of the proposed enzyme-bound intermediates have the capacity for the release of free NO, but that the extent of this release would be determined to a large degree by the flux of nitrite and reductants through the sequence. This in turn would be open to influence from interactions between the denitrification apparatus of the particular bacterial species or strain and important environmental influences such as the external redox potential, oxygen tension, the availability of particular electron donors and the nitrite concentration. Thus the results of the different studies may critically depend upon the precise conditions to which the bacteria were exposed during the experiment. In particular, the presence of traces of oxygen in the reaction mixtures of all of the more recent bacterial nitrosation studies may have significantly affected the balance of products of denitrification in favour of free NO formation, in contrast to those studies carried out under more strict anaerobiosis, where evidence for free NO formation was not obtained.

Despite some remaining uncertainty about the precise nature of the nitrosating agent elaborated by particular denitrifiers, it is clear that amongst this group of organisms are those with the greatest potential for NNC formation. Indeed, model calculations (Leach *et al.* 1987a) suggest that N-nitrosation, particularly by denitrifying bacteria, could elevate N-nitroso compound concentrations in the achlorhydric stomach very significantly above those likely to be formed in the normal acid stomach. The composition of the bacterial flora, then, at any infected or colonized site within the body will be an important factor not only in determining the conversion of nitrate to nitrite (see above) but also in determining the ultimate production of N-nitroso compounds, since N-nitrosation activity is very variable between species and strains and potentially very prone to environmental modulation.

The N-nitrosation reactions mediated by macrophages and presumably those other mammalian cell types which produce NO are not, strictly speaking, the outcome of nitrite metabolism in the body. However, since they are likely to occur in association with endogenous nitrate biosynthesis it is important to consider their relevance. The substrate providing the N-nitrosating agent in these cases is L-arginine. In the case of macrophages, metabolism of L-arginine to NO only follows immune stimulation, during which a specific enzyme involved in this process is elaborated. The NO which is generated is presumably rapidly oxidized by molecular oxygen to NO_2; this then undergoes a range of further reactions, producing N_2O_3, N_2O_4, NO_2^- and NO_3^-. Any nitrosatable substrates present will be nitrosated by some of this N_2O_3 and N_2O_4 before it has reacted with surrounding water molecules to form NO_2^- and NO_3^- (Miwa *et al.* 1987). This will result in the local formation of potentially genotoxic NNC at the site of inflammatory response and potential tissue damage.

7.6 CONCLUSIONS

Under normal circumstances, nitrate exposure would appear to be unavoidable. The extent to which different populations and groups are exposed can differ quite dramatically, however, depending on factors such as the type of diet, water nitrate content, soil type, agricultural practices, etc. Control of the extent of nitrate exposure of some populations and individuals, where there may be an associated health risk, should be possible.

While it is clear that endogenous nitrate biosynthesis provides a further source of nitrate exposure, the extent of this process in man in health is not, in the opinion of the authors, clear. Several estimates suggest a contribution as large as, or greater than, that from the normal diet. While it is clear that its contribution to overall exposure as measured by nitrate excretion will become most evident on low-nitrate diets, it is not clear why a similar level has not been seen to underlie urinary nitrate excretion in the wide range of other studies which have used higher dietary nitrate exposures. This point is raised in the review by Hartman (1982). Certainly in all of our studies it is impossible to reconcile the levels of urinary nitrate excretion of our volunteers, whilst ingesting very carefully controlled diets, with a high level of endogenous nitrate biosynthesis taking place.

If endogenous nitrate biosynthesis accounts for as much as 1 mmol, or more, a day, this would be reflected as a daily urinary excretion of greater than >0.6 mmol/day over and above that originating from the 60% of dietary nitrate also appearing in urine. In common with several other workers we fail to find such a high underlying rate of excretion. The level of nitrate excretion which cannot be accounted for by immediately preceding dietary nitrate intake in our studies is about 0.2–0.4 mmol/day and represents the maximum endogenous synthesis in these subjects. It remains possible that low-nitrate diets themselves have the effect of stimulating those processes which give rise to endogenous synthesis and that the high levels of endogenous nitrate biosynthesis measured under these conditions are an artifact of experimental design since very-low-nitrate diets are not ordinarily encountered in the field. It seems likely at present that, in health, human nitrate exposure occurs largely through the diet. However, in several disease states the contribution from endogenous synthesis may become temporarily dominant. An important feature of this source of nitrate is that a primary reaction product is nitric oxide, a molecule which, on oxidation and further reaction, can give rise to potent nitrosating agents. Thus individuals whose immune systems are appropriately stimulated may be transiently exposed to an important group of genotoxins, N-nitroso compounds.

Ingested nitrate and presumably endogenously formed nitrate are distributed to a greater or lesser extent throughout different body compartments. In several of these, bacterial infection or colonization can result in the conversion of nitrate to nitrite by the action of bacterial nitrate reductases. Thus in the mouth, the achlorhydric stomach and the infected urinary bladder, the bacterial metabolism of nitrate provides a dominant source of nitrite exposure, since dietary nitrite levels are extremely low by comparison with the levels found in these bacterially colonized sites. In health, approximately 55% of both dietary and endogenously formed nitrate are excreted in urine. Of the 45% of nitrate not recovered in urine, about 20% appears to be lost to bacterial metabolism in the lower GI tract, with the nitrate entering the lower bowel not by bulk flow but by secretion across the gut wall. The further 20% or so of nitrate not recovered in urine is thought to be lost as a consequence of 'mammalian processes', perhaps including losses through perspiration, tears, etc. However, it must be pointed out that bacterial nitrate metabolism in the mouth is likely to be a further source of nitrate loss which has yet to be taken into account quantitatively. The development of a nitrate-reducing bacterial flora in the stomach in association with achlorhydria can lead to further significant losses (up to 23% on average) of nitrate in the human body, in which case the average urinary recovery of dietary nitrate may become closer to 33% rather than 55%.

Nitrate and nitrite concentrations in all body fluids fluctuate dramatically throughout the course of the day. Such fluctuations can be attributable to intrinsic circadian patterns in metabolism; the sporadic nature and variable levels of nitrate ingestion; environmental factors influencing the nitrate reductase activity of the endogenous flora; factors influencing urinary and salivary flow; and possibly also periods of exercise. Thus only when expressed over reasonably extended time periods does the relationship between urinary nitrate levels and nitrate ingestion become reliable. The analyses of single 'spot' samples of any body fluid by contrast are unlikely to provide reliable estimates of dietary nitrate exposure. Indeed, only the analysis of 24 h urinary nitrate excretion has been advocated as a reliable method of determining the nitrate exposure of a population. Under these circumstances it has been calculated that because of the underlying urinary nitrate which is excreted independently of immediate nitrate ingestion (probably including endogenous synthesis) 24 h urinary nitrate values have to be adjusted to account for the 70% of apparent recovery of previous 24 h nitrate ingestion. Calculations based on a correction factor for 55% recovery, which is the actual urinary recovery of discrete dietary nitrate challenges, would overestimate *dietary* nitrate intake since the contributions from 'other sources' (i.e. endogenous nitrate synthesis and body pool leaching) would also inevitably become included unless these were first subtracted. From our studies of healthy UK populations the following relationships would apply:

(a) Dietary nitrate intake = (Urinary nitrate−0.22)×1/0.55
= (Urinary nitrate− 0.22)×1.8182

or

(b) Dietary nitrate intake = Urinary nitrate×1/0.7
= Urinary nitrate×1.428

The application of the one correction factor of 1.428 in (b) above, to make allowance for an apparent 70% recovery of dietary nitrate intake, will tend to overestimate lower nitrate intakes and underestimate higher ones, but is approximately accurate over the range of intakes 1 to 2 mmol (Table 7.5). Application of the relationship given above in (a) will provide much better estimates of nitrate intake generally and particularly when outside the range 1–2 mmol.

In assessing the *dietary* nitrate exposure of a population through urinary nitrate determinations it is clearly going to be important to exclude two groups of individuals from this part of an epidemiological analysis. That is those individuals having or likely to have achlorhydria and those with infections which promote an immune response. The former will be excreting rather less than 55% of dietary nitrate and the latter may well be excreting large amounts of endogenously produced nitrate.

The formation of nitrite from nitrate within bacterially colonized sites in the body may have in itself significant health implications, particularly from the point of view of nitrite formation in the stomach and the upper small intestine of infants, enhancing the risk of methaemoglobinaemia. A further risk from nitrate metabolism within the body is N-nitroso compound formation. The largest proportion of N-nitroso compounds is carcinogens and/or mutagens, and some are extremely potent. Their formation within the body can result from at least three distinct processes. Firstly they may arise during the formation of NO, nitrite and nitrate by macrophages and other cell types. Secondly they have been shown to be formed in

Table 7.5 — Nitrate intake, urinary recovery and calculation of nitrate intake from urinary excretion: based on two models

Nitrate intake	Urinary nitrate $Y=0.55X+0.22$	Calculated ingested 1 $X=1.8182(Y-0.22)$	Calculated ingested 2 $X=1.428Y$
0.2	0.33	0.2	0.47
0.4	0.44	0.4	0.63
0.6	0.55	0.6	0.78
0.8	0.66	0.8	0.94
0.9	0.71	0.9	1.02
1.0	0.77	1.0	1.10
1.2	0.88	1.2	1.26
1.5	1.04	1.5	1.49
1.8	1.21	1.8	1.73
2.0	1.32	2.0	1.80
2.2	1.43	2.2	2.04
2.5	1.59	2.5	2.28
3.0	1.87	3.0	2.67

the normal acid stomach as a consequence of the acid-catalysed reactions of nitrite, the nitrite being derived from nitrate, which during salivary recirculation is reduced to nitrite by the oral flora. Ordinarily, however, the nitrite concentrations in the acid stomach are low and limit the extent of NNC formation. Finally they may be formed by bacterially catalysed reactions at sites within the body which are or have become colonized by microorganisms, such as the mouth, the achlorhydric stomach and the infected urinary bladder. In this case the microbial flora is not only responsible for the N-nitrosation reaction but also for supplying increased levels of the substrate, nitrite, *in situ*.

All of these reactions producing N-nitroso compounds will clearly be influenced by a range of factors, such as the pH, the level of nitrate ingestion and the presence of catalysts and inhibitors of N-nitrosation. Those sites where the role of bacteria has the opportunity of becoming dominant in determining the extent of N-nitroso compound formation (for example the achlorhydric stomach and infected urinary bladder) will also have another set of variables modulating NNC exposure. One will be the species and strain composition of the bacterial flora in relation to nitrate and nitrite reductase and N-nitrosation activity. And these in themselves will be open to significant modulation from such factors as the external redox potential and electron donor availability within each bodily site, and the outcome of competition between the different bacterial species colonizing each site. Given the complexity and interaction of factors which would be expected to modulate NNC formation from nitrate, both host related and microbial flora associated, it remains unlikely that a simple relationship between nitrate exposure and the cancer risk of endogenous NNC formation exists.

REFERENCES

Adam, B., Schlag, P., Friedl, P., Preussmann, R. and Eisenbrand, G. (1989). Proline is not useful as a chemical probe to measure nitrosation in the gastro-intestinal tract of patients with gastric disorders characterised by anacidic conditions. *Gut* **30**, 1068–1075.

Alsobrook, A. J., Duplessis, L. S., Harrington, J. C., Nunn, A. J. and Nunn, J. R. (1974). Nitrosamines in the human vaginal vault. In: *N-nitroso Compounds in the Environment.* Eds P. Bogovski and E. Walker, IARC Scientific Publication 57, IARC, Lyon, pp. 197–201.

Archer, M. C. (1984). Catalysis and inhibition of N-nitrosation reactions. In: *N-nitroso Compounds: Occurrence, Biological Effects and Relevance to Human Cancer*, Eds I. K. O'Neill, R. C. von Borstel, J. E. Long, C. T. Miller and H. Bartsch. IARC Scientific Publication 57, IARC, Lyon, pp. 263–274.

Averill, B. A. and Tiedje, J. M. (1982). The chemical mechanism of microbial denitrification. *FEBS Lett.* **138**, 8–12.

Bartholomew, B. and Hill, M. J. (1984). The pharmacology of dietary nitrate and the origin of urinary nitrate. *Fd. Chem. Toxicol.* **22**, 789–795.

Billiar, T. R., Curran, R. D., Stuehr, D. J., Ferrari, F. K. and Simmons, R. L. (1989). Evidence that activation of Kupffer cells results in production of L-arginine metabolites that release cell-associated iron and inhibit hepatocyte protein synthesis. *Surgery* **106**, 364–372.

Bingham, S. (1987). The dietary assessment of individuals: methods, accuracy, new techniques and recommendations. *Nutrition Abs. and Revs.* **57**, 705–742.

Boss, P. M. J., Van Den Brandt, P. A., Wedel, M. and Ockhuizen, T. H. (1988). The reproducibility of the conversion of nitrate to nitrite in human saliva after a nitrate load. *Fd. Chem. Toxicol.* **26**, 93–97.

Brimblecombe, P. and Stedman, D. H. (1982). Historical evidence for a dramatic increase in the nitrate component in acid rain. *Nature* **298**, 460–462.

Calmels, S., Ohshima, H., Vincent, P., Gounot, A.-M. and Bartsch, H. (1985). Screening of micro-organisms for nitrosation catalysis at pH 7 and kinetic studies on nitrosamine formation from secondary amines by *E. coli* strains. *Carcinogenesis* **6**, 911–915.

Calmels, S., Ohshima, H. and Bartsch, H. (1988). Nitrosamine formation by denitrifying and non-denitrifying bacteria: implication of nitrite reductase and nitrate reductase in nitrosation catalysis. *J. Gen. Microbiol.* **134**, 221–226.

Caygill, C. P. J., Bartholomew, B. and Hill, M. J. (1986). The relation between drinking water nitrate and total nitrate intake. *Aqua.* **2**, 94–97.

Caygill, C. P. J., Leach, S. A., Kirkham, J. S., Northfield, T. C. Hall, C. N. and Hill, M. J. (1987). Gastric hypochlorhydria as a risk factor for gastric and other cancers. In: *The Relevance of N-nitroso Compounds to Human Cancer: Exposures and Mechanisms.* Eds H. Bartsch, I. O'Neill and R. Schulte-Hermann, *IARC Scientific Publication* 84, 524–526.

Central Water Planning Unit (1977). Nitrate and water resources. Central Water Planning Unit, Reading.

Challis, B. C.and Challis, J. A. (1982). N-Nitrosamines and N-nitrosoimines. In: *The Chemistry of Amino, Nitroso and Nitro-compounds and their Derivatives*, Ed. S. Patai, Wiley, New York, p. 1151.

Chilvers, C., Inskip, H., Caygill, C. P. J., Bartholomew, B., Fraser, P. and Hill, M. (1984). Survey of dietary nitrate in well-water users. *Int. J. Epidemiol.* **13**, 324–331.

Correa, P. (1985). Precancerous lesions of the stomach. In: *Diet and Human Carcinogenesis*, Eds J. V. Joosens, M. J. Hill and J. Geboers, ECP Symposium/2, Elsevier, Amsterdam, p. 109.

Dolby, J. M., Webster, A. D. B., Borriello, S. P., Barclay, F. E., Bartholomew, B. and Hill, M. J. (1984). Bacterial colonisation and nitrite concentration in the achlorhydric stomach of patients with primary hypogammaglobulinaemia and classical pernicious anaemia. *Scand. J. Gastroenterol.* **19**, 105–110.

Douglass, M. L., Kabacoff, B. L., Anderson, G. A. and Cheng, M. C. (1978). The chemistry of nitrosamine formation, inhibition and destruction. *J. Soc. Cosmet. Chem.* **29**, 581–606.

Drasar, B. S., Shiner, M. and McCleod, G. M. (1964). Studies on the intestinal flora. I. The bacterial flora of the gastrointestinal tract in healthy and achlorhydric persons. *Gastroenterology* **56**, 71–79.

Dull, B. J. and Hotchkiss, J. H. (1984). Activated oxygen and mammalian nitrate biosynthesis. *Carcinogenesis* **5**, 1161–1164.

Eisenbrand, G., Spiegelhalder, B. and Preussmann, R. (1980). Nitrate and nitrite in saliva. *Oncology* **37**, 227–231.

Forman, F., Al-Dabbagh, S. and Doll, R. (1985). Nitrates, nitrites and gastric cancer in Great Britain. *Nature* **313**, 620–625.

Forsythe, S. J. and Cole, J. A. (1987). Nitrite accumulation during anaerobic nitrate reduction by binary suspensions of bacteria isolated from the achlorhydric stomach. *J. Gen. Microbiol.* **133**, 1845–1849.

Forsythe, S. J., Dolby, J. M., Webster, A. D. B. and Cole, J. A. (1988). Nitrate- and nitrite-reducing bacteria in the achlorhydric stomach. *J. Med. Microbiol.* **25**, 253–259.

Garber, E. A. E. and Hollocher, T. C. (1981). ^{15}N Tracer studies on the role of NO in denitrification. *J. Biol. Chem.* **256**, 5459–5465.

Garber, E. A. E. and Hollocher, T. C. (1982). ^{15}N, ^{18}O Tracer studies on the activation of nitrite by denitrifying bacteria. *J. Biol. Chem.* **257**, 8091–8097.

Goretski, J. and Hollocher, T. C. (1988). Trapping of nitric oxide produced during denitrification by extracellular hemoglobin. *J. Biol. Chem.* **223**, 2316–2323.

Granli, T., Dahl, R., Brodin, P. and Bockman, O. C. (1989). Nitrate and nitrite concentrations in human saliva: variations with salivary flow rate. *Fd. Chem. Toxicol.* **27**, 675–680.

Green, L. C, Wagner, D. A., Ruiz de Luzuriaga, K., Istfan, N., Young, V. R. and Tannenbaum, S. R. (1981). Nitrate biosynthesis in man. *Proc. Nat. Acad. Sci.* **678**, 7764–7768.

Greene, I. and Hiatt, E. P. (1953). Behaviour of the nitrate ion in the dog. *Am. J. Physiol.* **176**, 463–467.

Hartman, P. E. (1982). Nitrates and nitrites: ingestion, pharmacodynamics and toxicology. In: *Chemical Mutagens*, Volume 7, Eds F. J. de Serres and A. Hollander, Plenum, pp. 211–293.

Hartman, P. E. (1983). Nitrate/nitrite ingestion and gastric cancer mortality. *Environ. Mutagen.* **5**, 111–121.

Hayashi, N., Watanabe, K., Ishiwata, H., Mizushiro, H., Tanimura, A. and Kurata, H. (1978). Fate of nitrate and nitrite in saliva and blood of monkey administered orally sodium nitrate solution and microflora of oral cavity of the monkey. *J. Food Hyg. Soc. Japan.* **19**, 392–400.

Hegesh, E. and Shiloah, J. (1982). Blood nitrates and infantile methaemoglobinaemia. *Clin. Chim. Acta.* **125**, 107–115.

Hibbs, J. B., Read, J. R., Taintor, R. and Vavrin, Z. (1987). Macrophage cytotoxicity: role of L-arginine deaminase and imino nitrogen oxidation to nitrite. *Science* **235**, 473–476.

Hicks, R. M., Walters, C. L., Elsebai, I., El Aasser, A.-b., El Merzabani, M. and Gough, T. A. (1977). Demonstration of nitrosamines in human urine: preliminary observations on possible etiology of bladder cancer in association with chronic urinary tract infection. *Proc. Roy. Soc. Med.* **70**, 413–417.

Hill, M. J. (1979). *In vivo* bacterial nitrosation and its possible role in human cancer, In: *Naturally occurring carcinogens–mutagens and modulators of carcinogenesis.* Ed. E. C. Miller. Japan Sci. Soc. Press, Tokyo, 229–240.

Hochstein, L. I. and Tomlinson, G. A. (1988). The enzymes associated with denitrification. *Ann. Rev. Microbiol.* **42**, 231–261.

Ishiwata, H., Boriboon, P., Nakamura, Y., Harada, M., Tanimura, A. and Ishidate, M. (1975). Studies on *in vivo* formation of N-nitroso (II). Changes in nitrite and nitrate concentrations in human saliva after ingestion of vegetables or sodium nitrate. *J. Food Hyg. Soc. Japan.* **16**, 19–24.

Ishiwata, H. and Tanimura, A. (1982). Ingestion and excretion of nitrate in humans and experimental animals. In: *Environmental Mutagens and Carcinogens.* Eds T. Sugimura, S. Kondo and H. Takebe. University of Tokyo Press, Tokyo.

Iyengar, R., Stuehr, D. J. and Marletta, M. A. (1987). Macrophage synthesis of nitrite, nitrate and N-nitrosamines: precursors and the role of the respiratory burst. *Proc. Nat. Acad. Sci.* **84**, 6369–6373.

Ji, X.-B. and Hollocher, T. C. (1988a). Reduction of nitrite to nitric oxide by enteric bacteria. *Biochem. Biophys. Res. Comm.* **157**, 106-108.

Ji X-B, and Hollocher, T. C. (1988b). Mechanism of nitrosation of 2,3-diaminonaphthalene by *Escherichia coli*: enzymatic production of NO followed by O_2-dependent chemical nitrosation. *App. Env. Microbiol.* **54**, 1791–1794.

Knight, T. M., Forman, D., Al-Dabbagh, S. A. and Doll, R. (1987). Estimation of dietary intake of nitrate and nitrite in Great Britain. *Fd. Chem. Toxicol.* **25**, 277–285.

Knowles, R. (1982). Denitrification. *Microbiol. Rev.* **46**, 43–70.

Knowles, R. G., Palacios, M., Palmer, R. M. J. and Moncada, S. (1989). Formation of nitric oxide forms L-arginine in the central nervous system: a transduction mechanism for stimulation of the soluble guanylate cyclase. *Proc. Nat. Acad. Sci.* **86**, 5159–5162.

Leach, S. A. (1988). Mechanisms of endogenous N-nitrosation. In: *Nitrosamines: Toxicology and Microbiology*. Ed. M. J. Hill, Ellis Horwood, Chichester, p. 69.

Leach, S. A., Challis, B. C., Cook, A. R., Hill, M. J. and Thompson, M. H. (1985). Bacterial catalysis of the N-nitrosation of secondary amines. *Biochem. Soc. Trans.* **13**, 381–382.

Leach, S. A., Thompson, M. H. and Hill, M. J. (1987a). Bacterially catalysed N-nitrosation reactions and their relative importance in the human stomach. *Carcinogenesis* **8**, 1907–1912.

Leach, S. A., Cook, A. R., Challis, B. C., Hill, M. J., Thompson, M. H. (1987b). Bacterially mediated N-nitrosation reactions and endogenous formation of N-nitroso compounds. In: *The Relevance of N-nitroso Compounds to Human Cancer: Exposure and Mechanisms*. Eds H. Bartsch, I. O'Neill and R. Schulte-Herman IARC Scientific Publication 84, IARC, Lyon, p. 396.

Leach, S. A., Mackerness, C. W., Thompson, M. and Hill, M. (1988). The role of bacteria and N-nitroso compounds in gastric cancer—development of an *in vitro* model. *Cancer Lett.* **39** (Suppl), S40.

Leach, S. A. and Hill, M. J. (1990). The effects of the sustained and profound inhibition of gastric acid production on gut flora. In: Ed. J. B. Elder, *Profound Gastric Acid Suppression: A Long-term Safety Risk? Research and Clinical Forums*, Vol. 12, pp. 139–151.

Leaf, C. D., Vecchio, A. J., Roe, D. A. and Hotchkiss, J. H. (1987). Influence of ascorbic acid dose on N-nitroso compound formation in humans. *Carcinogenesis* **8**, 791–795.

Leaf, C. D., Wishnok, J. S. and Tannenbaum, S. R. (1989). L-Arginine is a precursor to nitrate biosynthesis in humans. *Biochem. Biophys. Res. Comm.* **163** , 1032–1037.

Leaf, C. D., Wishnok, J. S. Hurley, J. P., Rosenbald, W. D., Fox, J. G. and Tannenbaum, S. R. (1990). Nitrate biosynthesis in rats, ferrets and humans. Precursor studies with L-arginine. *Carcinogenesis* **11**, 855–858.

Licht, W. R., Schultz, D. S., Fox, J. G., Tannenbaum, S. R. and Deen, W. M. (1986). Mechanisms of nitrite loss from the stomach. *Carcinogenesis* **7**, 1681–1687.

MAFF; Steering Group on food Surveillance (1987). *Nitrate, Nitrite and N-nitroso Compounds in Food*. Food Surveillance Paper No. 20, HMSO, London.

Malagelada, J.-R., Robertson, J. S., Brown, M. L., Remington, M., Duenes, J., Thonforde, G. M. and Carryer, P. W. (1984). Intestinal transit of solid and liquid components of a meal in health. *Gastroenterol* **87**, 1255–1263.

Marletta, M. A. (1988). Mammalian synthesis of nitrite, nitrate, nitric oxide, and N-nitrosating agents. *Chem. Res. Toxicol.* **1**, 250–257.

Marletta, M. A. (1989). Nitric oxide: biosynthesis and biological significance. *TIBS* **14**, 488–492.

Marletta, M. A., Yoon, P. S., Iyengar, R., Leaf, C. D. and Wishnok, J. S. (1988). Macrophage oxidation of L-arginine to nitrate and nitrite: nitric oxide is an intermediate. *Biochemistry* **27**, 8706–8711.

Mayer, E. A., Thomson, J. B., Jehn, D., Reedy, T., Elashoff, J., Deveny, C. and Meyer, J. H. (1984). Gastric emptying and sieving of solid food and pancreatic and biliary secretions after solid meals in patients with nonresective ulcer surgery. *Gastroenterol* **87**, 1264–1271.

McCall, T. B., Boughton-Smith, N. K., Palmer, R. M. J., Wittle, B. J. R. and Moncada, S. (1989). Synthesis of nitric oxide from L-arginine by neutrophils. *Biochem. J.* **261**, 293–296.

McDermid, A. S., McKee, A. S., Ellwood, D. C. and Marsh, P. D. (1986). The effect of lowering the pH on the composition and metabolism of a community of nine oral bacteria grown in a chemostat. *J. Gen. Microbiol.* **132**, 1205–1214.

Meyrick-Thomas, J., Misiewicz, J. J., Cook, A. R., Hill, M. J., Smith, P. L. R., Walters, C. L., Forster, J. K., Martin, L. E. and Woodings, D. F. (1987). Effects of one year's treatment with ranitidine and of truncal vagotomy on gastric contents. *Gut* **28**, 726–738.

Milton-Thompson, G. J., Ahmet, Z., Lightfoot, N. F., Hunt, R. H., Barnard, J., Brimblecombe, R. W., Moore, P. J., Bavin P. M. G., Darkin, D. W. and Viney, N. (1982). Intragastric acidity, bacteria, nitrite and N-nitroso compounds, before, during and after Cimetidine treatment. *Lancet* **i**, 1091–1095.

Mitchell, H. H., Schonle, H. A. and Grindley, H. S. (1916). The origins of nitrate in urine. *J. Biol. Chem.* **24**, 461–490.

Miwa, M., Stuehr, D. J., Marletta, M. A., Wishnok, J. S. and Tannenbaum, S. R. (1987). Nitrosation of amines by stimulated macrophages. *Carcinogenesis* **8**, 955–958.

Moncada, S., Radomski, M. W. and Palmer, R. M. J. (1988). Endothelium-derived relaxing factor: identification as nitric oxide and role in the control of vascular tone and platelet function. *Biochem. Pharmacol.* **37**, 2495–2501.

Mueller, R. L., Hagel, H.-J., Greim, G., Ruppin, H. and Domschke, W. (1983). The endogenous synthesis of cancerogenic N-nitroso compounds: bacterial flora and nitrite formation in the healthy human stomach. *Zbl. Hyg. I. Abt. Orig. B.* **178**, 297–315.

Mueller, R. L., Hagel, H.-J., Greim, G., Ruppin, H. and Domschke, W. (1984). The dynamics of the endogenous bacterial nitrite formation in the stomach. *Zbl. Hyg. I. Abt. Orig. B.* **179**, 381–396.

Mueller, R. L. and Henninger, H. (1985). The gastric mucous gel-an endogenous nitrate reservoir. *Zbl. Bakt. b*, 181.

Mueller, R. L., Hagel, H.-J., Greim, G., Ruppin, H. and Domschke, W. (1986). Nitrate and nitrite in normal gastric juice. Precursors of endogenous N-nitroso compound synthesis. *Oncology* **43** 50–53.

Ohshima, H., Calmels, S., Pignatelli, B., Vincent, P. and Bartsch, H. (1987). N-nitrosamine formation in urinary-tract infection. In: *The Relevance of N-nitroso Compounds to Human Cancer: Exposures and Mechanisms*, Eds H. Bartsch, I. O'Neill and R. Schulte-Hermann, IARC Scientific Publication 84, pp. 384–390.

Packer, P. J. (1989). Metabolism and pharmacology of nitrate in relation to human cancer. CNAA, Ph.D. Thesis.

Packer, P. J., Leach, S. A., Randell, P. R. and Hill, M. J. (1988). Further evaluation of urinary nitrate as a measure of nitrate exposure in relation to human cancer. *Cancer Letters* **39** (Supp), S37.

Packer, P. J., Leach, S. A., Duncan, S. N., Thompson, M. H. and Hill, M. J. (1989). The effect of different sources of nitrate exposure on urinary nitrate recovery in humans and its relevance to the methods of estimating nitrate exposure in epidemiological studies. *Carcinogenesis* **10**, 1989–1994.

Packer, P. J., Van Acker, B., Reed, P. I., Haines, K., Thompson, M. H., Hill, M. J. and Leach, S. A. (1990). The effect of gastric achlorhydria on the urinary recovery of nitrate in man: relevance to urinary nitrate as a measure of dietary nitrate exposure. *Carcinogenesis* **11**, 1373–1376.

Palmer, R. M. J., Ferrige, A. G. and Moncada, S. (1987). Nitric oxide release accounts for the biological activity of endothelium-derived relaxing factor. *Nature* **327**, 524–526.

Payne, W. J. (1973). Reduction of nitrogenous oxides by micro-organisms. *Bacteriol. Rev.* **37**, 409–452.

Preussmann, R. and Tricker, A. R. (1988). Endogenous nitrosamine formation and nitrate burden in relation to gastric cancer epidemiology. In: *Gastric carcinogenesis*, Eds P. I. Reed and M. J. Hill, ECP Symposium/5, Elsevier, Amsterdam, p. 147.

Radomski, J. L., Palmiri, C. and Hearn, W. L. (1978). Concentration of nitrate in normal human urine and the effect of nitrogen ingestion. *Toxicol. Appl. Pharm.* **45**, 63–68.

Ralt, D., Wishnok, J. S., Fitts, R. and Tannenbaum, S. R. (1988). Bacterial catalysis of nitrosation: involvement of the *nar* operon of *Escherichia coli. J. Bacteriol.* **170**, 359–364.

Ruddell, W. S. J., Bone, E. S., Hill, M. J., Blendis, L. M. and Walters, C. L. (1976). Gastric juice nitrite. The risk factor for cancer in the hypochlorhydric stomach. *Lancet* **ii**, 1037–1039.

Ruddell, W. S. J., Bone, E. S., Hill, M. J. and Walters, C. L. (1978). Pathogenesis of gastric cancer in pernicious anaemia patients. *Lancet* **1**, 521–522.

Saul, R. L., Kabir, S. H., Cohen, Z., Bruce, W. R. and Archer, M. C. (1981). Nitrate formation in rats exposed to nitrogen dioxide. *Cancer Res.* **41**, 2280–2283.

Saul, R. L. and Archer, M. C. (1984). Oxidation of ammonia and hydroxylamine to nitrate in the rat and *in vitro*. *Carcinogenesis* **5** 77–81.

Schultz, D. S., Deen, W. M., Karel, S. F., Wagner, D. A. and Tannenbaum, S. R. (1985). Pharmacokinetics of nitrate in humans: role of gastrointestinal absorption and metabolism. *Carcinogenesis* **6**, 847–852.

Schulz, K.-R. von, Scheibe, J., Diener, W., Fischer, G., Namaschk, A., Thu, M. P. and Wettig, K. (1990). Endogene nitrat synthese bei ausgewahlten infektionskrankheiten. *Z. Klin. Med.* **45**, 925–927.

Shuker, D. E. G. (1988). The chemistry of N-nitrosation. In: *Nitrosamines: Toxicology and Microbiology*. Ed. M. J. Hill, Ellis Horwood, Chichester, p. 48.

Spiegelhalder, B., Eisenbrand, G. and Preussman, R. (1976). Influence of dietary nitrate on nitrite content of human saliva: possible implications to *in vivo* formation of N-nitroso compounds. *Fd. Cosmet. Toxicol.* **14**, 545–548.

Statutory Instruments: The preservatives in food (Amendment) regulations (1962, 1979 and 1982), HMSO, London.

Stephany, R. W. and Schuller, P. L. (1980). Daily intakes of nitrate, nitrite and N-nitrosamines in the Netherlands using the duplicate portion sampling technique. *Oncology* **37** 203–210.

Stewart, V. (1988). Nitrate respiration in relation to facultative metabolism in Enterobacteria. *Microbiol. Rev.* **52**, 190–232.

Stuehr, D. J. and Marletta, M. A. (1985). Mammalian nitrate biosynthesis: mouse macrophages produce nitrate and nitrite in response to *Escherichia coli* lipopolysaccharide. *Proc. Nat. Acad. Sci.* **82**, 7738–7742.

Stuehr, D. J. and Marletta, M. A. (1987). Synthesis on nitrate and nitrite in macrophage cell lines. *Cancer Res.* **47** 5590–5594.

Tannenbaum, S. R. (1979). Nitrate and nitrite: origins in humans. *Science* **205**, 1333–1335.

Tannenbaum, S. R., Sinsky, A. J., Weisman, M. and Bishop, W. (1974). Nitrite in human saliva, its possible relationship to nitrosamine formation. *J. Nat. Cancer Inst.* **53**, 79–84.

Tannenbaum, S. R., Weisman, M. and Fett, D. (1976). The effect of nitrate intake on nitrite formation in human saliva. *Fd. Chem. Toxicol.* **14**, 549–552.

Tannenbaum, S. R., Fett, D., Young, V. R., Land, P. D. and Bruce, W. R. (1978a). Nitrate and nitrite are formed by endogenous synthesis in the human intestine. *Science* **200**, 1487–1489.

Tannenbaum, S. R., Archer, M. C., Wishnok, J. S. and Bishop, W. W. (1978b). Nitrosamine formation in human saliva. *J. Nat. Cancer. Inst.* **60**, 251–253.

Tannenbaum, S. R., Moran, D., Rand, W., Cuello, C. and Correa, P. (1979). Gastric cancer in Columbia IV: nitrite and other ions in gastric contents of residents from a high risk region. *J. Nat. Cancer Inst.* **69**, 9–12.

Thayer, J. R., Chasko, J. H., Swartz, K. A. and Parks, N. J. (1982). Gut reactions of radioactive nitrite after intratracheal administration in mice. *Science* **217**, 151–153.

Theilade, E. (1989). Factors controlling the microflora of the healthy mouth. In: *Human Microbial Ecology.* Eds M. J. Hill and P. D. Marsh, CRC Press, Boca Raton, Florida, p. 1.

Tricker, A. R., Mostafa, M. H., Speigelhalder, B. and Preussmann, R. (1989). Urinary excretion of nitrate, nitrite and N-nitroso compounds in Schistosomiasis and Bilharzia bladder cancer patients. *Carcinogenesis* **10**, 547–552.

US National Academy of Sciences (1981). *The Health Effects of Nitrate, Nitrite and N-nitroso Compounds.* National Academy Press, Washington DC.

Wagner, D. A. and Tannenbaum, S. R. (1982). Enhancement of nitrate biosynthesis by *E. coli* lipopolysaccharide. In: *Nitrosamines in Human Cancer.* Ed. P. N. Magee, Banbury Report No. 12, Cold Spring Harbor, New York, pp. 437–443.

Wagner, D. A., Schultz, D. S., Deen, W. M., Young, V. R. and Tannenbaum, S. R. (1983). Metabolic fate of an oral dose of [^{15}N]-labelled nitrate in humans: effect

of diet supplementation with ascorbic acid. *Cancer Res.* **43**, 1921–1925.
Walters, C. L. and Smith, P. L. R. (1981). The effect of water-borne nitrate on salivary nitrate. *Fd. Chem. Toxicol.* **19**, 297–310.
Wang, F. C., Cassens, R. G. and Hoekstra, W. G. (1981). Fate of ingested ^{15}N-labelled nitrate and nitrite in the rat. *J. Food Sci.* **46**, 745–748.
Wettig, K., Dobberkau, H.-J. and Flentje, F. (1990). Elevated endogenous nitrate synthesis associated with Giardiasis. *J. Hyg. Epidemiol. Microbiol. Immunol.* **34**, 69–72.
Whelan, M. (1935). The nitrate content of animal tissues and fate of ingested nitrate. *Biochem. J.* **29**, 782–787.
White, J. W. (1975). Relative significance of dietary sources of nitrate and nitrite. *J. Agric. Food Chem.* **23**, 886–891.
White, R. J. (1983). Nitrate in British waters. *Aqua.* **2**, 51–57.
Witter, J. P. and Balish, E. (1979). Distribution and metabolism of ingested nitrate and nitrite in germ free and conventional flora rats. *Appl. Environ. Microbiol.* **38**, 861–869.
Witter, J. P., Gatley, S. J. and Balish, E. (1979a). Nitrate and nitrite: origins in humans. *Science* **205**, 1335–1337.
Witter, J. P., Balish, E. and Gatley, S. J. (1979b). Distribution and metabolism of ingested nitrate and nitrite in germ free and conventional-flora rats. *Appl. Environ. Microbiol.* **38** 870–878.

8

Nitrates and nitrites from food and water in relation to human disease

M. J. Hill

8.1 INTRODUCTION

Until 1945 there was little interest in the amount of nitrate or nitrite in the human diet. There is little evidence of acute toxicity associated with nitrate ingestion whilst nitrite, although known to cause dilation of blood vessels (and so depression of blood pressure) and production of methaemoglobin, was disregarded as a general hazard.

The health aspects can be subdivided into acute toxicity and the effects of chronic dosage. The chronic toxicity of nitrate is due to its reduction to nitrite in the body.

8.2 ACUTE EFFECTS

8.2.1 Acute toxicity of nitrite

The major source of toxicity of nitrite is the formation of methaemoglobin (Met-Hb) from oxyhaemoglobin (oxy-Hb). Haemoglobin (Hb) is an iron-containing complex present in erythrocytes with the principal role of oxygen transport. It combines with oxygen to form oxy-Hb, which is red and which dissociates readily to release oxygen to the tissues. A range of oxidizing compounds react with oxy-Hb to form Met-Hb (which is chocolate brown). Met-Hb has greatly reduced oxygen binding; that oxygen which is present is very strongly bound, so that Met-Hb is not available to the essential alternating oxygenation and deoxygenation involved in the transport of oxygen from the lungs to the tissues. When Met-Hb is less than 10% of the total Hb it is usually asymptomatic; above 10% it causes 'cyanosis' in which first the lips then the skin takes a bluish-grey hue and the blood is noticeably chocolate brown in colour. Early symptoms are of lassitude, but when a high proportion of the haemoglobin is in the form of Met-Hb the affected person goes into coma and may die; the generally accepted lethal level of Met-Hb is 60%.

Oxy-Hb is converted *in vitro* to Met-Hb by nitrite according to the equation (Kosaka *et al.* 1979).

$$4HbO_2+4NO_2+4H \rightarrow 4MHb+4NO_3^-+O_2+2H_2O$$

This equation applies when there is a large excess of nitrite over haemoglobin (at least eightfold, according to Doyle *et al.* (1985)); further, the reaction is an extremely complex one (Doyle *et al.* 1985, Kosaka and Uozumi 1986), and attempts to use the reaction to calculate toxic doses of nitrite (Winton *et al.* 1970) have not been successful or reliable.

Ingestion of sodium nitrite in large amounts can be fatal, the lethal dose being 2–9 g as $NaNO_2$ (WHO, 1985); the figure of 4 g $NaNO_2$ is widely accepted. This appears to be based on incidents such as one in 1946 in which 71 persons in Liepzig consumed soup that had been mistakenly 'salted' using curing brine rich in $NaNO_2$ (Schultze and Scheibe 1948). Of these, 37 required hospitalization and 7 died. In humans the first symptoms of oral nitrite poisoning occur 15–45 minutes after dosage (Schmidt *et al.* 1949, Aquanno *et al.* 1981).

Kiese and Weger (1969) carried out studies of the intravenous toxicity of nitrite. When volunteers were given 4 mg $NaNO_2$ per kg body weight they developed maximal methaemoglobin levels of 7%; 12 mg/kg gave 25% methaemoglobin. Assuming a linear dose–response relationship and that 60% methaemoglobin is lethal, these results indicate an i.v. lethal dose of 1.5–2.5 g for an average adult of 60–70 kg body weight. Sodium nitrite has occasionally been used in suicide.

8.2.2 Acute toxicity of nitrate

The pharmacology of oral nitrate is dealt with in detail elsewhere in this volume. Large doses of ammonium nitrate have been administered to adults with no major symptoms. For example, Ellen *et al.* (1982) gave 0.15 g/kg body weight of ammonium nitrate as single oral doses to 12 volunteers; one developed diarrhoea after 12 h and one vomited after 12 min, but none had any elevation of methaemoglobin levels or other haematological effects. Ammonium nitrate is regularly administered at 10 g/day for 4–5 days as a diuretic, with no other significant symptomatic effects. Bruijns (1982) reported on the treatment of 268 patients treated with 2–9 g/day ammonium nitrate for 3 weeks to 2 years; only two cases of methaemoglobinaemia were reported.

Amundsen (1947) reported an outbreak of nitrate poisoning in which 15 persons who had consumed vegetables accidentally treated with sodium nitrate became ill and 15 died of acute methaemoglobinaemia. The mean lethal dose was approximately 15 g; however, the authors suggested that the vehicle of the nitrate load (yellow turnips) would be relatively poorly digested, leading to slow gastric emptying and so greater reduction to nitrite. In consequence the toxic effects can be ascribed to those of nitrite rather than of nitrate *per se,* and had the nitrate been administered as part of a more readily digestible meal the lethal dose would have been higher.

8.3 CHRONIC EFFECTS

There are no reports of chronic toxic effects of nitrate that are not related to the effect of the product of bacterial reduction, nitrite. The major health hazards cited in the literature are infant methaemoglobinaemia and N-nitroso-related carcinogenesis. In addition there have been reports of nitrate ingestion being associated with a number of other less well documented infant diseases such as congenital abnormalities.

8.4 INFANT METHAEMOGLOBINAEMIA

8.4.1 Clinical aspects and diagnosis

Cyanosis is a rare disease in which the affected child develops a brownish-blue discoloration of the skin; this is apparent first in the lips, then spreads to the fingers and toes, and may eventually cover the whole body. Cyanosis is the result of oxygen starvation in the tissues and can be caused by cardiovascular or respiratory problems, or by methaemoglobinaemia; in the latter case the disease may be the result of known drug effects, but if this is not so, and the disease has been associated with intestinal upset, it is known as idiopathic or enterogenous cyanosis or methaemoglobinaemia.

Infant methaemoglobinaemia, or 'blue-baby syndrome', is known to be caused by a range of compounds, such as the early sulphonamides, aniline and derivatives, nitrophenol, acetanilide, phenacetin and potassium chlorate. Such compounds have long been known and their avoidance has, in general, not proved to be difficult. In addition, a number of nitrates and nitrites have also been implicated. The first report was of bismuth subnitrate, which had been used in the treatment of infant enteritis until the risk of infant methaemoglobin formation was recognized by Roe (1933).

The role of nitrate and nitrite in methaemoglobin formation has long been known. Gamgee (1868) showed that amylnitrate, when added to blood, caused methaemoglobin formation. Administration of large doses of nitrates to animals was shown to cause methaemoglobin formation by Binz and Geringer (1901); Stokvis (1902) described the formation of methaemoglobinaemia as the cause of enterogenous cyanosis. In 1928, Barker and O'Hare described a case of cyanosis caused by treatment with ammonium nitrate (used in the treatment of diuresis). Roe (1933) reported the death from cyanosis of a young infant treated with bismuth subnitrate for enteritis. He is credited in the literature survey by Walton (1951) with being the first to associate nitrate exposure to serious or fatal infant methaemoglobinaemia. In 1934, Chen *et al.* (1934) demonstrated the production of methaemoglobinaemia following intravenous injection of sodium nitrite.

This background set the scene for the seminal paper by Comly (1945) which is usually credited as the first report of the role of nitrate in drinking water in the causation of infant methaemoglobinaemia.

8.4.2 Early evidence of a role for nitrate

In 1945, Comly reported two cases of 'a previously unrecognised condition which may be confused with congenital heart disease' and which could 'occur anywhere in rural areas where well water is used in infant feeding'. The first infant was admitted to hospital on a series of occasions with cyanosis, was put on to a range of alternative formula feeds, the only common factor in which was the water used to make up the feed. All known causes of cyanosis were ruled out. The well water was analysed and shown to contain 619 mg/litre of nitrate. When the baby was put on to a half-skimmed acidified milk formula that did not require any added water she thrived. The second case followed a similar course; the well water used to make up the feed contained 388 mg/litre of nitrate. In both cases, the immediate treatment of the cyanosis was by the administration of methylene blue, and recurrence was prevented by avoidance of the well water. Comly then reported five further cases of cyanosis in Iowa State, all associated with high levels of nitrate in the drinking water. Anecdotally he reported five further cases from a paediatrician elsewhere in the state, and noted a report of a further five cases of idiopathic cyanosis reported at a meeting in 1941. In all cases the infants had intestinal disturbances as a prelude to the cyanosis. All had been consuming formula feeds made up with the local well water.

Comly concluded, after a review of the literature and of his evidence, that the enterogenous cyanosis seen in his patients and elsewhere in Iowa was due to the ingestion of large amounts of nitrate (from the well water) in the formula feed. The intestinal upset permitted the establishment of bacteria in the infant gut able to reduce nitrate to nitrite; this latter ion was then absorbed and was the cause of the methaemoglobinaemia. Comly further noted that Zobell and Meyer (1932) had demonstrated *in vitro* the production of nitrate reductase by many of the organisms found in the infant gastrointestinal tract.

This paper was followed by a host of reports, all confirming the association between well water rich in nitrate, the used formula feeding, intestinal disease and cyanosis. Enterogenous cyanosis is virtually unknown in breast-fed infants, and is very rare in infants more than a few months old. Lecks (1950) reviewed 31 literature cases, all of which had in common (a) age less than 90 days, (b) born into a rural community, (c) fed on formula feed made up with boiled well water containing 66–619 mg/l nitrate. The main symptom of cyanosis was in all cases progressive, and started approximately one month after the onset of use of well water for making up the feed. Cornblath and Hartman (1948) extended the observations and reported that cyanosis only occurs following ingestion of high-nitrate water in infants with a gastric pH>4 and with an intestinal flora risk in nitrate-reducing bacteria. The high gastric pH was postulated to permit the outgrowth of the nitrate-reducing bacteria, thereby increasing the nitrite load on the infant. Ewing and Mayon-White (1951) reported the first cases from the UK, reviewed the literature, and stressed that the effect was cumulative. Ferrant (1946) reported two cases from Belgium. Walton (1951) carried out a useful survey of the literature to that date. By 1972, Shearer *et al.* estimated that 1500 to 2000 cases had occurred worldwide since the first reports by Comly in 1945. since then there have been further reports from elsewhere. For example, Super *et al.* (1981) reported further cases from Namibia.

8.4.3 Proposed mechanism

In utero the infant human is germ-free and is only contaminated on the outside during delivery. Bacteria are ingested from the environment and, because there is little gastric acid secretion for the first months of life, the infant intestine is unprotected from such organisms. The absence of a resident bacterial flora means that there is no colonization-resistance to such organisms. During the first few days, therefore, the infant gut is colonized by a range of strictly aerobic environmental organisms (*Pseudomonas* spp, *Staphylococcus* spp, *Bacillus* spp) which would not normally be able to thrive in the intestine. Breast-fed infants then develop a flora which by day 30 is dominated by *Bifidobacterium* spp, *Lactobacillus* spp, *Streptococcus* spp and *Enterococcus* spp (Roberts 1988). The source of this flora is not known but is thought to include the breast during suckling. Such a flora would maintain the gut anaerobic, and so prevent the growth of the strictly aerobic environmental organism previously seen simply by removal of the essential oxygen. Further, the organisms are all lactic-acid-producing species and, because human milk has low buffering capacity, consequently is able to acidify the gastric contents to below pH 4. This is thought to assist in preventing colonization by the enteric pathogens and the putrefactive organisms (*Salmonella* spp, *Bacteroides* spp). The mechanisms of this colonization resistance are not understood (although there are many hypotheses, reviewed by Hill *et al.* (1986) and Roberts (1988)) but the simplicity and the type of flora are established. Thus, in summary, in breast-fed infants the flora is a simple lactic acid flora dominated by *Bifidobacterium* spp (Table 8.1) and with low nitrate reductase activity. Further, they are fed on breast milk with no direct exposure to the local water supply. Finally, breast-fed infants have a low risk of enteric infection (thought by Comly (1945) to result in a damaged mucosa more 'permeable' to nitrite). Such infants do not develop enterogenous cyanosis. By the time they are weaned to solid foods (made up, usually in tap water) they have begun to secrete gastric acid and so have an established gastric acid barrier and consequently an upper intestine relatively free of bacteria. Such infants do not have a nitrate reducing flora in that part of the gut exposed to nitrate- and so are not at serious risk of methaemoglobinaemia. This is summarized in Table 8.2.

In contrast, the formula-fed infant consumed a commercial formula feed that was rich in protein, low in lactose and high in buffering capacity. Such infants develop a flora (Table 8.1) rich in putrefactive organisms (to utilize the protein) and which is able to colonize because the gastric pH remains high. Such a flora is rich in nitrate reductase but in infant enteritis there is an outgrowth of intestinal pathogens, particularly in the small bowel, most of which are potent producers of nitrate reductase. As in the breast-fed infants, it takes a few weeks to establish this flora, dominated by coliforms, *Bacteroides* spp and *Clostridium* spp (in addition to large

Table 8.1 — The bacterial flora of breast-fed compared with formula-fed infants. Data are expressed as mean $\log_{10}$ (bacterial count per gram faeces; figures in parentheses are the percentage of infants carrying the organism

	Breast fed	Formula fed
Bifidobacterium spp	9 (>90)	9 (>90)
Streptococcus spp	6 (>90%)	6 (70–90%)
Lactobacillus spp	8 (>90%)	6 (70–90%)
Coliforms	5 (50–80%)	9 (>90%)
Bacteroides spp	4 (<30%)	8 (80–100%)
Clostridium spp	4 (<30%)	8 (80–100%)

Table 8.2 — The factors thought to predispose to cyanosis and their prevalence pre- and post-weaning

Factor	Pre-weaning		Post-weaning
	Breast fed	Bottle fed	
Gastric acid secretion	–	–	++
Gastric pH	4	7	2
Nitrate-reducing bacterial in upper gut	±	+++	–
Risk of enteritis (and mucosal damage)	±	++	–
Erythrocyte nitrate reductase	–		++

numbers of relatively inoffensive *Bifidobacterium* spp). The comparative development of the infant flora is illustrated in Table 8.3.

Table 8.3 — Development of the infant flora in breast-fed and bottle-fed infants

	Breast fed	Formula fed
Flora at 7 days	Unstable mixture of environmental organisms	Unstable mixture of environmental organisms
Key characteristics of the milk diet	Low protein, high lactose, low buffering	High protein, low lactose, high buffering
Flora at day 30	Dominated by Bifido-bacteria streptococci (i.e. lactic acid bacteria)	Mixture of Bifids, coliforms, Clostridia and Bacteroides (i.e. putrefactive flora)
Gastric pH at day 30	~4	~7
Flora at day 90	Similar to day 30	Similar to day 30
Post-weaning flora	Dominated by Bifids and Bacteroides; Streptococci, coliforms and Clostridia minor components	As for breast fed

Infants are particularly susceptible to methaemoglobinaemia for a number of reasons. Because of their very high growth rate during the early weeks of life, their food and fluid intake per kilogram body weight is very high and consequently the nitrate load is very high. The lack of gastric acidity results in the upper gastro-intestinal tract being heavily colonized by bacteria able to reduce nitrate to nitrite. In older infants with developed gastric acidity the upper intestinal tract is relatively clear of bacteria; although the lower intestine is very heavily colonized, this is not relevant in health because the nitrate will be absorbed from the small bowel before it can be exposed to high nitrate reductase activity. Finally, infant erythrocytes have no nitrate reductase activity and so the methaemoglobin remains active; in older infants the erythrocyte enzyme degrades the nitrohaemoglobin thereby releasing the haemoglobin to carry out its oxygen transport function.

8.4.4 The current position

Methaemoglobinaemia is not a notifiable disease in the UK and so its incidence is unknown; only 10 cases of methaemoglobinaemia have been reported in the UK since 1945 (and only one death) but this almost certainly underestimates the situation.

However, the disease has been notifiable in Hungary since 1968. In the first five years there were 883 cases reported. Between 1975–1977 there were 190 cases in four counties (total population 1.7 million) and six deaths; in almost all cases the babies were less than 90 days old and consumed water with more than the WHO limit of 22.6 mg/l of nitrate nitrogen. From 1976 to 1982 in the whole of Hungary there were 1353 cases of methaemoglobinaemia, with 21 deaths. Following the instigation of measures to supply drinking water with less than 9 mg/l nitrate nitrogen there has been a definite decrease in the number of cases reported annually since 1979 (WHO 1985). Thus, although there have been few reported cases in the UK (and only occasional cases in Scandinavia and the Netherlands) it is clear from the Hungarian data that methaemoglobinaemia can be an important public health problem in rural populations with well-water supplies.

Shearer *et al.* (1972) showed that all infants exposed to nitrate, whether symptomatic or not, had detectable levels of methaemoglobin, but the level increased with increasing nitrate concentrations in the water and was higher in infants with intestinal disease than in those with no gut disease. A number of studies from the Soviet Union suggested that the problem of methaemoglobinaemia might not be limited to young infants less than 90 days old but might affect school age children (Diskalenko 1968, Kapanadze 1961, Sublotin 1961, Petukhov and Ivanov 1970). In addition to reporting elevated levels of Met-Hb in schoolchildren Petukhov and Ivanov (1970) also reported clinical symptoms associated with consumption of well water containing 23.7 mg/l nitrate nitrogen such as lassitude, slowing of conditioned motor reflexes. This clinical effect caused considerable worry at the time but has not been repeated in western populations. Craun *et al.* (1981) in a study of young American schoolchildren, reported no elevated level of methaemoglobin in young infants (up to 8 years drinking water containing 22 to 111 mg per litre of nitrate nitrogen; this was discussed in relation to the then-current EPA limit for nitrate in water of 10 mg per litre nitrate nitrogen and to the 'protective effect' of age. This latter is now thought to be due to the combined effect of the development of a functioning gastric acid barrier (thereby reducing the flora of the upper intestine and so decreasing the nitrite load) and the development of erythrocyte enzymes which degrade nitrite. Lower Met-Hb levels are invariably reported from areas where the vitamin C intake is high and it is possible that this explains the discrepancy between the Russian and western experience. Super *et al.* studied Met-Hb levels in Namibian infants and found a close correlation with nitrate intake from water; as in the Russian studies they found no relation with age and this was postulated to be due to the lack of vitamin C in the diet of these African children as illustrated in Table 8.4. The protective effects of gastric acidity and of Met-Hb reductase can be demonstrated by the known increased risk of methaemoglobinaemia in adults with reduced gastric acidity and in individuals who have a hereditary lack of the erythrocyte enzyme.

Table 8.4 — The interrelation between vitamin C intake, nitrate intake and the frequency of elevated methaemoglobin levels (data adapted from Super *et al.* (1981))

Vitamin C supplementation	Nitrate intake†	Methaemoglobin values (percent) <2	>3
None	Low	81	12
	High	52	37
Infrequent	Low	79	12
	High	54	35
Daily	Low	73	12
	High	94	6

†Nitrate intake graded high if the drinking water nitrate contained more than 20 mg/l nitrate nitrogen.

8.4.5 Conclusions

1. Infantile methaemoglobinaemia is now rare in northern and western Europe and in urban areas of North America, but data from Hungary suggest that the disease is probably relatively common in the less affluent, more rural areas of eastern and southern Europe and in the rest of the world. Although it has a multifactorial etiology the nitrate concentration in the diet, and particularly in the water used to make up infant foods, is an important contributing factor, the effect of which is amplified when the well water is bacterially contaminated or when the infant has intestinal disease.
2. Clinical symptomatic disease is limited mainly to babies less than 90 days old in populations with an inadequate vitamin C intake, but may be seen in older children when the vitamin C intake is low, and in adults with low gastric acid secretion.
3. In the earlier studies, when enteritis was much more common, and in the Hungarian studies, almost all cases occurred in babies consuming feeds made up with water containing more than 22.6 mg/l of nitrate nitrogen (although 7% of Hungarian cases occurred in babies with drinking water between 11 and 22.6 mg/l).
4. It is my opinion that, although enteritis is much less common than previously, nevertheless the nitrate limits have to be set on the assumption that outbreaks will

inevitably occur even when the public health measures are at their best. The current limits appear to be well justified.

8.5 NITRATE AND HUMAN CANCER

The hypothesis that nitrate exposure might be implicated in the causation of human cancer stemmed from the realization that N-nitroso compounds form a family of highly carcinogenic compounds that can be formed *in vivo* in humans as well as in foods treated with nitrite or nitrate. To appreciate the cause for the concern, it is necessary to review the chemistry, the formation and the carcinogenicity of N-nitroso compounds. Such a review should lead to an appreciation of the magnitude of the potential hazard, but epidemiology is then necessary to determine the magnitude of the actual hazard. This latter will be achieved, however, only if the epidemiology is performed in the light of the best information from other sciences.

8.5.1 N-nitroso compounds

N-nitroso compounds are formed by the action of nitrous acid on a suitable secondary nitrogen group; if the parent nitrogen group is a secondary amine then the product is an N-nitrosamine, amides give rise to N-nitrosamides, ureas give N-nitrosoureas, etc. The production of such compounds was of little interest to non-specialists until it was reported, as part of a study of the hepatotoxicity suffered by workers in the rubber industry, that dimethyl nitrosamine was a potent carcinogen in rodents (Magee and Barnes 1956). This led to a spate of reports of the structure–activity relationships of N-nitrosamine carcinogenicity, which showed that a wide range of such compounds were carcinogenic. Further, the compounds proved to be organotropic carcinogens, with the tumour site being dependent on the nature of the parent amine and of the test animal (Table 8.5). In 1967, Magee and Barnes published an important review from which it became clear that cancer could be induced at almost any site in rodents by suitable choice of the nitrosamine and the test animal. Further, since the test compound could be introduced by any route this obviated the problem of tissue damage caused at the injection site, obscuring early histological signs of malignant change. For these reasons, the use of N-nitrosamines in the laboratory study of carcinogenesis became common, and this led inevitably to speculation that such compounds might also be implicated in human carcinogenesis.

Although the N-nitrosamines are target organ-specific and cause tumours at sites distant from that of their introduction into the body, N-nitrosamides and N-nitrosoureas are locally acting and cause tumours only at their site of introduction. The N-nitrosamines are not direct-acting mutagens but need activation by microsomal enzymes before they are mutagenic in the Salmonella mutagenesis assay; they also need to be activated in the body, but this process, which leads to the organotropism of these compounds, is little understood. In consequence we cannot predict with confidence the target organ of any N-nitrosamine in any test animal. In contrast, the N-nitrosoamides or N-nitrosoureas are direct-acting mutagens which require no activation and so are able to act at the site of introduction into the body.

The chemistry of N-nitroso compounds was reviewed recently by Shuker (1988) and their carcinogenicity and toxicology by Rowland (1988). In the literature,

Table 8.5 — Target organ specificity of various N-nitroso compounds (data from various sources, principally from Magee and Barnes (1967))

N-nitroso compound	Animal	Target organ
Dimethylnitrosamine	Rat	Kidney
	Hamster	Liver
Dibutylnitrosamine	Rat	Bladder
	Mouse	Bladder
N-butylethylnitrosamine	Mouse	Forestomach
	Rat	Oesophagus
N-nitrosopiperidine	Rat	Oesophagus
	Hamster	Lung
N-nitrosomorpholine	Rat	Liver
	Hamster	Lung
N-nitrosopiperazine	Mouse	Lung
	Rat	Oesophagus
N-nitrosoamides	Rat	Local
N-nitrosoureas	Rat	Local

N-nitroso compounds are classified loosely in various ways. They may be classified as volatile and non-volatile; the volatile nitroso compounds are the N-nitrosamines in which the alkyl side chains are simple unsubstituted hydrocarbon chains or with simple aromatic substituents, whilst the non-volatile compounds include the N-nitrosoamides, N-nitrosoureas or N-nitrosopeptides, etc. They may be described as 'total N-nitroso compounds' (TNC) or 'total apparent N-nitroso compounds' (TANC) (Massey 1988, Walters *et al.* 1980, Reed *et al.* 1981); in this case they have usually been analysed by a group reaction which assays the number of N-nitroso bonds. This assay gives no information on the type of N-nitroso compound or on its relative carcinogenicity. Total volatile N-nitroso compounds include only those that can be assayed by gas chromatography. In some studies the non-volatile N-nitroso compounds refer only to the N-nitroso amino acids found in normal human urine (e.g. Tricker *et al.* 1989). These differences in definitions need to be taken into account when comparing quantitatively the assays in different studies.

8.5.2 Production of N-nitroso compounds *in vivo*

N-nitroso compounds can be formed by the action of nitrous acid on a suitable nitrogen compound, and the pH-optimum of the reaction varies with the substrate (Table 8.6), as does the stability of the product. Since the pH-optima are usually below 4, the only site in the body in which they are likely to be produced in large amounts by the acid-catalysed reaction is the acid stomach, and this has been demonstrated (Fine *et al.* 1977, Ohshima and Bartsch 1981). However, in 1968, Sander demonstrated that the N-nitrosation reaction could be catalysed by bacterial action at neutral pH values, and using nitrate as well as nitrite as the substrate. This

Table 8.6 — The pH-optimum of the N-nitrosation of various classes of nitrogen compound

Parent nitrogen compound	pH-optimum of N-nitrosation
Secondary amine	3.0–3.5
Tertiary amines	3.0–3.4
Quaternary ammonium salts	5.6
Amino acids (proline)	2.5
Amides, alkylureas, etc.	<2

work was confirmed and extended by numerous groups (Table 8.7), but early studies failed to distinguish between catalysis by a bacterial enzyme and catalysis of the acid-catalysed reaction by the conditions created by bacterial metabolism. The problem was resolved by the careful studies of Calmels *et al.* (1985) and by Leach *et al.* (1985), reviewed by Leach (1988), which showed unambiguously the existence of a bacterial enzyme which catalysed the N-nitrosation of secondary amines. Further, all of the early work in Table 8.7 was carried out using enterobacteria which have low nitrosating activity; Leach *et al.* (1987) demonstrated the existence of a group of 'super-nitrosator' species with enzyme activity which exceeded that of the enterobacteria by a factor of 100–10 000. Leach *et al.* (1987) calculated that the amount of N-nitroso compound likely to be formed in the achlorhydric stomach by bacterial action exceeded that likely to be produced at pH 2 by the acid-catalysed reaction. This established the prime importance of the bacterially catalysed reaction, which was confirmed by analysis of gastric juice by Walters *et al.* (1980), Reed *et al.* (1981) and Pignatelli *et al.* (1987). The last-mentioned study resolved a conflict between the two major procedures for the analysis of TANC — those of Walters *et al.* (1980) and of Bavin *et al.* (1982). Bacterial catalysis of N-nitrosation has been reviewed recently by Leach (1988).

This demonstrated that N-nitroso compounds will be formed whenever bacteria, nitrosatable nitrogen compounds and nitrate/nitrite are present together. The amount will depend in part on the concentration of nitrate/nitrite present. The importance of such bacterially formed N-nitroso compounds can only be demonstrated by epidemiological methods. In early studies, the epidemiologists concentrated on the relation between nitrate exposure and the overall human cancer risk.

8.5.3 Epidemiology of nitrate exposure and human cancer

The first such study was carried out by Zaldivar (1970), who reported that there was a correlation between nitrate exposure in the 17 regions of Chile and the incidence of oesophageal cancer. In later studies, Zaldivar and Robinson (1973) showed a correlation between nitrate fertilizer usage and gastric cancer incidence; this was supported by further studies from Armijo and Coulson (1975), Zaldivar and Wetterstrand (1975) and Zaldivar (1977). It is known that the nitrate content of vegetables increases with fertilizer usage, and so this should be a measure of nitrate

Table 8.7 — Evidence that bacteria are able to catalyse the N-nitrosation reaction

Reference[†]	Observation
Sander (1968)	First demonstrated N-nitrosation of a secondary amine at neutral pH in the presence of *E. coli* but not in their absence.
Hawksworth and Hill (1971)	Confirmed the above and extended the range of organisms studied. Showed the need for growing cultures.
Klubes *et al.* (1972)	Repeated the above and showed that part of the activity was destroyed by autoclaving (and so could be enzymic) but part of the activity was resistant (and so was non-enzymic)
Brooks *et al.* (1972)	First reported N-nitrosation *in vivo* in urine by *Proteus* spp.
Coloe and Heywood (1976)	Showed a wide variation in activity between *E. coli* strains; studied the kinetics of the reaction.
Tannenbaum *et al.* (1978)	Studied N-nitrosation in saliva *in vitro* and showed that live bacteria were needed.
Ruddell *et al.* (1978)	Demonstrated N-nitrosation by bacteria in gastric juice.
Kunisaka and Hayashi (1979)	Studied N-nitrosation by resting cells of *E. coli* and concluded that the reaction was enzymic.
Leach *et al.* (1985)	Confirmed that the reaction is enzymic; showed that denitrifiers are very much more active than non-denitrifiers; reported that the pH-optimum was between 7 and 8.

[†]For details of references, see Hill (1986).

exposure. Hill *et al.* (1973) argued that, although in normal circumstances more than 80% of the dietary nitrate exposure is from food, the total intake and the proportion from water increase as the nitrate concentration in the drinking water increases. When the nitrate content of the water reaches 100 mg/l nitrate (as NO_3) the water should then account for 70% of the total nitrate intake, and the total nitrate per day should have increased by approximately 150–200% (Table 8.8); this assumption was confirmed in a large study in East Anglia by Caygill *et al.* (1987). Thus, *for the purpose of carrying out an epidemiological study,* it is advisable to compare the cancer risk in areas with different nitrate levels in the drinking water. Such a preliminary study (Hill *et al.* 1973) suggested an excess risk of gastric cancer associated with high nitrate exposure; these results have been supported by many studies (Table 8.9) and contradicted by many others.

All of these studies have been subject to heavy criticism (e.g. Fraser *et al.* 1980, Fraser and Chilvers 1981); the value of these early observations was that they began a debate on the causation of gastric cancer.

Table 8.8 — The sources of nitrate in the diet and the effect of varying the nitrate content of drinking water on total nitrate intake

	Amount consumed (g per week)	Mean nitrate content (mg/kg)	Intake of nitrate per week	
			Water has 15 mg/l	Water has 100 mg/l
Processed meats	450 g	200	90 (15%)	90 (6%)
Vegetables (excluding potatoes)	700 g	500	350 (60%)	350 (24%)
Water	10 l		150 (25%)	1000 (70%)
		Total	590	1440

Table 8.9 — Some studies of the relation between drinking water nitrate and gastric cancer risk in populations. For details and references, see Hill (1984)

Country	Prevalence of gastric atrophy	Source of the high nitrate exposure
Chile	High	Vegetables (due to high-nitrate fertilizer usage)
China	High	Vegetables (due to low molybdenum in the soil)
Colombia	High	{Vegetables (due to high
Japan	High	{nitrate in volcanic soil)
Hungary	Intermediate	Drinking water†
Italy	Intermediate	Drinking water†
Israel	Low	Drinking water†
UK	Low	Drinking water
Denmark	Low	Drinking water†

†'High nitrate' was more than 100 mg/l of nitrate; in studies comparing low values of drinking water nitrate, no relation to gastric cancer was seen (because water is then only a minor contributor to the nitrate load; see Table 8.8).

8.5.4 Etiology of gastric cancer

Gastric cancer undoubtedly has a multifactorial etiology, since the stomach is the first resting place of all dietary carcinogens. In consequence, it is probably unrealistic to attempt to determine 'the causation' of the disease. Nevertheless there have been many epidemiological studies, particularly of the role of diet and of familial (genetic) factors. In general, these studies only ever find weak correlations, the most common of which are listed in Table 8.10. A major reason for this lack of clarity is that gastric

Table 8.10 — Epidemiology of gastric cancer

Factor	Comments
Genetic factors	Blood groups very weakly correlated. Small excess risk in first-order relatives
Environmental factors	
Non-dietary	Dusty work (e.g. coal mining) Smoking (particularly in Japanese studies) Socioeconomic status (inverse) Sex (M/F=approx 2)
Dietary	Grain products Root vegetables/starchy foods Salad vegetables (protective) Smoked or salted foods

cancer has been treated as a single disease; there have been many classifications of gastric cancers, but the most useful in studies of etiology has been that of Lauren (1965) which divided the disease into intestinal and diffuse types (with a third, and intermediate, type added by most users of the scheme). The major distinctions between the two have been reviewed recently by Day (1981) and by Filipe (1987). In the context of this chapter, the key observation was made by Lehtola (1978), who noted that the diffuse type was strongly associated with familial/genetic factors whilst the intestinal type was not associated with genetic predisposition. In contrast the intestinal-type disease is strongly associated with environmental/dietary factors (Correa 1983, Munoz and Asvall 1971). Thus, in case-control studies of the role of nitrate in gastric carcinogenesis only cases of intestinal-type disease should have been included, and all diffuse-type cases should have been excluded; no epidemiological study reported to date has done this and so, inevitably, the picture has been confused.

A further, and equally important, factor that has blurred most of the analyses carried out to date has been the fact that gastric cancer (like all other cancers) is a multistage process. A factor might be very important at a single stage and play no role in the other stages and, under such circumstances, may only be weakly associated epidemiologically with the overall process whilst nevertheless being of crucial importance to a key step. Correa *et al.* (1975) postulated a histopathological sequence for gastric carcinogenesis (Fig. 8.1) in which the first stage was gastric atrophy (GA), followed by chronic atrophic gastritis (CAG), then intestinal metaplasia (IM), epithelial dysplasia (ED) of increasing severity and finally gastric

Normal mucosa
↓
Gastric atrophy (GA)
↓
Atrophic gastritis (AG)
↓
Intestinal metaplasia (IM)
↓
Epithelial dysplasia (ED)
↓
Gastric carcinoma
(intestinal type)

Fig. 8.1 — The histopathological sequence in gastric cancer.

carcinoma (GC). They further hypothesized that N-nitroso compounds play an important role in the later stages of the disease. On the model of Correa *et al.* the major first step in the causation of the disease is GA, which results in a progressive decrease in the secretion of gastric acid until the gastric luminal pH can no longer be brought down to below 4 (the bactericidal level, according to the studies of, for example, Bartholomew *et al.* 1980). When this stage is reached, the gastric lumen is no longer a self-sterilizing system and a resident bacterial flora establishes in the stomach which is able to reduce nitrate to nitrite and then to catalyse the subsequent N-nitrosation reaction at a high rate if the flora includes 'supernitrosation'. This flora is then responsible for production of N-nitroso products from dietary nitrate and from (probably) endogenous nitrosatable nitrogen compounds; these N-nitroso compounds are hypothesized to cause the progression through intestinal metaplasia and increasingly severe dysplasia to gastric cancer.

If this hypothesis is true, then it inevitably follows that:

(a) All disease states or treatments that are associated with loss of gastric acidity will be associated with an excess risk of gastric cancer.
(b) All situations in which there is loss of gastric acidity will be accompanied by the establishment of a resident gastric bacterial flora rich in nitrate reductase and in N-nitrosating ability.
(c) In patients with decreased gastric acidity, the gastric juice bacterial flora and nitrite concentration should be correlated with the severity of epithelial dysplasia.
(d) If the combination of bacterial colonization, high-nitrate and nitrosatable nitrogen compounds can cause local cancers in the stomach then they should also cause local tumours if the same combination exists elsewhere in the body.
(e) Since it is unlikely that only locally acting N-nitroso compounds will be formed in the stomach, if sufficient N-nitroso compound is formed to cause local tumours there should also be high amounts of those acting at distant sites.

All of these predictions have been examined critically, and there is strong evidence to support each, as illustrated in Tables 8.11 to 8.15. In particular, Table 8.12 shows the strength of the relation between increased gastric juice nitrite and pH, whilst Table

8.13 shows the strength of the relation between gastric juice nitrite concentration and the severity of epithelial dysplasia of gastric mucosa. In the light of the strength of the evidence for the proposed mechanism, and the lack of contrary evidence, the role of nitrate/nitrite in causing the progression from intestinal metaplasia to gastric cancer cannot be disputed. However, the size of the risk has still to be determined.

Table 8.11 — The relation between gastric pH and risk of gastric cancer

Disease state	Risk of gastric cancer
Gastric atrophy	Increased
Chronic atrophic gastritis	Increased 2–3 fold
Intestinal metaplasia	Increased 2–4 fold
Pernicious anaemia	Increased 4–6 fold
Polyapartial gastrectomy	Increased 4–6 fold after 20-year latency
Hypogammaglobulinaemia	Increased 50 fold
Vagotomy	Increased 6–10 fold

8.6 RISK ASSESSMENT OF GASTRIC CANCER IN RELATION TO NITRATE INTAKE

On the proposed hypothesis already discussed above, there is no reason to postulate any risk of gastric cancer associated with high nitrate consumption in persons with normal gastric acidity. In persons with low gastric acidity there is a latency period of 20 years from the time of establishment of a resident bacterial flora before the increased risk becomes manifest (Caygill *et al.* 1986, Stahlsberg and Taksdal, 1971, Watt *et al.* 1984). Chen *et al.* (1990) the calculated risk of dysplasia in persons initially diagnosed with CAG (and loss of gastric acidity) and reported a risk ratio of more than 15 fold for gastric juice nitrite; Adam *et al.* (1989) showed that in achlorhydric patients the gastric juice nitrite level increased with nitrate intake. Although there are no solid data relating the two quantitatively, it would appear likely that the risk of gastric cancer (after a latent period of 20 years) in persons with no functioning gastric acid barrier will increase with the nitrate intake. The need is therefore to be able to assess the magnitude of the at-risk population.

Loss of gastric acidity may be the result of surgery, of specific genetically determined diseases, of natural ageing, of dietary deficiencies or of microbial infection. Those with iatrogenic loss of acidity are principally those treated for peptic ulcer; until 1980 most of these would have received surgery, but more recently the necessity for this has been largely obviated by the development of drugs able to

Table 8.12 — The relation between the gastric juice pH and the gastric juice concentration of bacteria and nitrite (for details and references, see Hill (1988))

Reference	Patient Group	Effect of increased pH	
		Bacterial flora	Nitrite
Drasar *et al.* (1969)	Various	Increased	
Ruddell *et al.* (1976)	Various	Increased	Increased
Ruddell *et al.* (1978)	PA and Controls	Increased	Increased
Jones *et al.* (1978)	PG		Increased
Walters *et al.* (1978)	Various	Increased	Increased
Schlag *et al.* (1980)	PG	Increased	Increased
Bartholomew *et al.* (1980)	PA and controls	Increased	Increased
Schumpelink *et al.* (1980)	PG	Increased	Increased
Reed *et al.* (1981)	Various	Increased	Increased
Muscroft *et al.* (1981)	Various	Increased	Increased
Watt *et al.* (1984)	Various	Increased	Increased
Keighley *et al.* (1984)	Gastric	Increased	Increased
Stockbrugger *et al.* (1984)	PA and control	Increased	Increased
Pounder *et al.* (1984)	Patients on omeprazole	Increased	Increased
Dolby *et al.* (1984)	Various	Increased	Increased
Caygill *et al.* (1984)	PG and control	Increased	Increased
Mortensen *et al.* (1984)	PG	Increased	Increased
Zhang *et al.* (1984)	Gastric atrophy	Increased	Increased
Stockbrugger *et al.* (1985)	Various	Increased	Increased
Cook *et al.* (1985)	Patients on ranitidine	Increased	Increased
Kyrtopoulos *et al.* (1985)	Various	Increased	Increased
Carboni *et al.* (1986)	Gastric surgery	Increased	Increased
Thomas *et al.* (1987)	Patients on	Increased	Increased
Savalgi *et al.* (1989)	PG	Increased	Increased

inhibit acid secretion in a more controlled way which avoids the loss of the gastric acid barrier. The gastric surgery patients are a readily identifiable group, as are the patients with genetically determined diseases (e.g. pernicious anaemia and hypogammaglobulinaemia). Estimating the size of the group who have lost their gastric acid barrier through natural ageing (and identifying the affected persons) is much more difficult. Vanzant *et al.* (1932) analysed records of 3242 persons with no recognizable gastric disease (Table 8.16) and found a surprisingly high incidence of achlorhydria at all ages, the incidence increasing linearly with age. Since Caygill *et al.*

Table 8.13 — The relationship between gastric juice analyses and the severity of dysplasia of the gastric mucosa (for details and references, see Hill (1991))

Reference	Patient group	Observation
Jones *et al.* (20)	Gastric surgery	Nitrite concentration correlated to severity of dysplasia
Watt *et al.* (21)	Mixed	Severity of dysplasia correlated to nitrite and bacterial concentration
Mortensen *et al.* (22)	Gastric surgery	Severity of dysplasia correlated to nitrite concentration
Keighley *et al.* (18)	Mixed	Risk of gastric cancer correlated with nitrite concentration
Savalgi *et al.* (23)	Gastric surgery	Severity of dysplasia correlated to nitrite concentration
Chen *et al.* (24)	Mixed	Severity of dysplasia correlated to nitrite concentration

Table 8.14 — The risk of local cancers in other sites in the body where bacteria nitrite and nitrosatable nitrogen compound coexist

Site	Observation
Urinary bladder	Excess risk of bladder cancer in urinary tract infection (Rodamski *et al.*, 1978)
	Excess risk of bladder cancer in patients with bacterial infection secondary to bladder bilharzia (Hicks *et al.* 1977)
	Excess risk of bladder cancer in paraplegia (Davies 1977)
Colon in urine diversion	Excess risk of cancer at the anastomotic site (Stewart *et al.* 1981)
Vagina	Excess risk of cervical cancer in women with *T. vaginalis* infection (Robertson *et al.* 1971)

Table 8.15 — The risk of cancer at various sites following gastric surgery for peptic ulcer[†]

Site	Excess risk of cancer after surgery	
	0–20 years	>20 years
Stomach	1.2	4.5
Large bowel	0.7	1.6
Biliary tract	1.9	8.6
Breast	0.5	4.0
Pancreas	0.7	3.8
Bladder	0.6	2.4

[†]From Caygill *et al.* (1987).

(1986) showed a latency of 20 years post loss of gastric acid before the risk of gastric cancer became above that in the general population, only those with achlorhydria before the age of 50 are at risk. On the basis of the data of Vanzant *et al.*, more than 10% of the population would be classified as 'at risk', the proportion being greater in women than in men. The risk of gastric achlorhydria is known to be related to diet (Gracey *et al.* 1977, Thomason *et al.* 1981) and it is undoubtedly true that the nutritional status of the general population has improved markedly since the date of the Vanzant study. Nevertheless it is likely that the prevalence of persons aged less than 50 with a resident gastric flora is likely to be at least 5% and may be up to 10% (since many more persons will have a gastric pH of more than 5 than have a pH as high as 7). Thus, the at-risk population cannot be regarded as insignificant.

Table 8.16 — The percentage of persons with gastric hypochlorhydria in relation to age (data from Vanzant *et al.* (1932))

Males	Males		Females	
	Number	Percentage	Number	Percentage
20–29	331	3.3	250	4.8
30–39	401	3.5	389	9.5
40–49	466	9.9	278	12.6
50–59	352	17.6	320	18.2
60–69	259	23.2	196	27.6

It remains to explain why, for example, East Anglia has a rising exposure to nitrate and a decreasing gastric cancer risk, and why there is conflict in the interpretation of the data from western European countries. In the study of Hill *et al.* (1973), the risk of gastric cancer in Worksop (where the drinking water used to contain 90 mg/l of nitrate) was compared with that in a number of towns with a similar socioeconomic status and similar major employment pattern (i.e. coal mining and iron/steel industry). It was observed (Table 8.17) that there was an excess risk of gastric cancer in Worksop males only in those aged more than 75 whilst in females the excess was detectable also in the 65–74-year-old band. Caygill *et al.* showed an excess risk of gastric cancer after a 20-year latency of 4.6 fold. The proportion of persons under 40 likely to have been achlorhydric would have been too small to have had a major effect on the gastric cancer incidence.

Table 8.17 — Death rates from gastric cancer in Worksop compared with control towns† (data from Hill *et al.* (1973), expressed as deaths per 100 000 persons per annum)

Age	Males		Females	
	Control towns	Worksop	Control towns	Worksop
35–44	0.3	0.8	0.3	0.6
45–64	5.3	5.3	2.3	2.5
65–74	20.2	22.2	8.9	15.2
⩾75	29.2	49.0	21.7	42.0

†Control towns were Sheffield, Wakefield, Chesterfield, Doncaster and Matlock, chosen because they matched Worksop for socioeconomic structure, major types of employment location, etc., but had low drinking water nitrate.

Under these circumstances we should not expect to see any excess number of cases in persons under 70 years of age (although the effect would be expected to be greater in females than in males, as in fact was observed). Within East Anglia there are grounds for at least suspecting that the prevalence of achlorhydria in young persons has decreased since the 1940s, as has been observed in, for example, Iceland or Finland. Thus the observed decrease in gastric cancer incidence in East Anglia during the last 40 years tells us more about the effect of improved nutrition on gastric acid secretion than on the role of nitrate in the later stages of gastric carcinogenesis.

A further cause of the disagreement between different groups of the data on the relation between nitrate and gastric cancer stems from the use by epidemiologists of truncated cancer rates, covering persons aged 35–64. The advantage of this is that the diagnoses are more reliable in persons less than 65 years old than in older persons (where the age of the patient leads clinicians to be less invasive in their diagnoses and where many of the sequelae of ageing can confuse the diagnosis). The major disadvantage is that by using truncated rates the age group most likely to be involved would be excluded from the analysis. Thus a falsely optimistic picture has been observed because the gastric cancer rate is in any case falling for unrelated reasons and because statistics tend to be used which reinforce the optimism by omitting the age group most likely to be affected.

8.7 NITRATE AND NON-GASTRIC CANCERS

The evidence relating nitrate to non-gastric cancers is dependent on the putative role for N-nitroso compounds in human carcinogenesis. The cancer sites fall into two groups:

(a) Those where locally formed N-nitroso compounds cause local tumours.
(b) Those where N-nitroso compounds formed in, for example, the stomach are active at distant sites.

8.7.1 Locally acting N-nitroso compounds

Sites in the body where bacteria, nitrate/nitrite and nitrosatable amino compounds occur together, other than the stomach, are the urinary bladder, saliva, the small bowel and large intestine. Of these, the incubation time and the concentrations of reactants are low in saliva and it has been difficult even to demonstrate the formation of N-nitroso compounds. Similarly, the small bowel has very low bacterial concentrations and rapid flow of luminal fluid, and so is an unlikely site for significant N-nitroso compound formation. The two sites of possible interest are the chronically infected urinary bladder and the colon.

8.7.1.1 Cancer of the urinary bladder and bladder infections

In western populations, symptomatic infections of the urinary bladder are common, affecting as many as 20% of the population at some time during the course of a year (Sinclair and Tuxford, 1971). These are of short duration, the infecting organism washing out. Most symptomatic infections involve a single infecting organism, usually a coliform organism (Savage *et al.* 1967) as illustrated in Table 8.18. Such organisms are potent producers of nitrate reductase but tend to be relatively poor producers of N-nitroso compounds in comparison with the 'supernitrosating' denitrifying species (Table 8.19) as demonstrated by Leach *et al.* 1985. *In vivo* production of N-nitroso compounds in the infected urinary bladder was first demonstrated in rats by Hawksworth and Hill (1974) and in humans by Brooks *et al.* (1972); this latter has been confirmed by, for example, Hicks *et al.* (1977) and by Tricker *et al.* (1989). Rodamski *et al.* (1978) reported that women with chronic urinary tract infections had an excess risk of bladder cancer.

The excess risk of local malignancy associated with urinary tract infections was not large, but would not be expected to be so on the nitrosamine hypothesis because the exposure to nitroso compounds is not large. In contrast, in patients with bilharzial infection of the bladder there is a secondary overgrowth of bacteria. The damage to the bladder wall that is caused by the bilharzial cysts provides a focus for bacterial colonization and so the transient infections that would in any case normally occur. Instead of repeated re-infection the patient gets repeated superinfection and rapidly acquires a urinary flora that is rich and complex and almost certainly includes some of the 'supernitrosating' strains. In consequence, in the urine of bilharzial patients there is a high concentration of N-nitroso compounds (Hicks *et al.* 1977, Tricker *et al.* 1989), and it has been postulated by Hicks *et al.* that the bladder cancer known to be commonly associated with bilharzial infection of the bladder is due to the N-nitroso compounds formed by the secondary bacterial infection rather than to a direct effect of the bilharzial cysts.

Table 8.18 — Organisms most commonly responsible for urinary tract infection

Bacterial species	Percentage
Coliforms	
Escherichia coli	80
Klebsiella spp	5–10
Proteus spp	5–10
Non-coliforms	
Staph. albus	<5
Pseudomonas spp	<2
Strep. faecalis	<2
Anaerobic bacteria	<2

Table 8.19 — The rate of N-nitrosation by various bacterial species (data from Leach *et al.* (1985))

Non-denitrifying species	Rate	Denitrifying species	Rate
Esch. coli strains	0–58	*Ps. aeruginosa*	660–12000
Streptococcus spp	0	*Neisseria* spp	240–1300
Veillonella spp	0	*Alc. faecalis*	2300
Clostridium spp	0	*B. licheniformis*	1620
Bifidobacterium spp	0		

Hicks *et al.* (1977) pointed out that in paraplegia and quadriplegia, complex chronic urinary tract infection is very common and it has been shown that here, too, there is an increased risk of bladder cancer (Davies 1977).

If the bladder cancer in these situations is caused by locally formed and locally acting N-nitroso compounds, then the risk of malignancy should increase with increasing nitrate intake. The only evidence for this is weak and comes from the study of Tricker *et al.* (1989); they found that in 27 bilharzial patients who had *not* progressed to malignancy, the urinary nitrate and nitrite (a good measure of nitrate intake, as noted in Chapter 7 of this book) was 148 mg/day, compared with 177 mg/day in 23 patients who had progressed to malignancy.

8.7.1.2 Cancer of the colon

There can be little quantitative evidence of the extent of N-nitroso compound formation in the normal human colon. Hawksworth and Hill (1971) showed that the vast majority of an oral dose of nitrate is absorbed from the upper gastrointestinal tract and very little reaches the colon. This was confirmed by Saul *et al.* (1981). The data given by Tannenbaum *et al.* (1978), which suggested that there were high levels of nitrate and nitrite in ileostomy fluid and, by inference, in the normal colon, have been very widely quoted but were withdrawn by Tannenbaum *et al.* (1981) on the grounds that the analytical methods used and the sample handling were inappropriate. In consequence there is no good evidence that significant amounts of this crucial substrate are present for local N-nitrosation in the colon.

In contrast, in patients with urinary diversion into the sigmoid colon, the high concentrations of nitrate and of nitrosatable amine present in urine are introduced into the colon, with its rich bacterial flora — the ideal conditions for bacterial N-nitrosation. Such diversion operations are carried out in patients from whom the urinary bladder has to be removed because of infection (tuberculosis, severe bilharzia, chronic severe cystitis), congenital abnormality, bladder malignancy, etc. The first successful procedure was to divert the urine into the sigmoid colon — ureterosigmoid anastomosis; this was first described by Stiles (1911) but suffered from problems of kidney infection due to retrograde progression of colonic bacteria up the ureters. In consequence an alternative procedure was adopted in which the ureters were conducted to an artificial bladder constructed from a length of ileum and having an exit through an abdominal stoma — referred to as an ileal loop. The first successful ileal loop was reported by Bricker (1950). Later, there was a valuable development in the ureterocolic anastomosis when the technique of 'tunnelling' the ureters under a mucosal flap to give a form of non-return valve was developed. The ileal mucosa is less conducive to this development and so the alternative colonic loop operation (in which the artificial bladder was constructed from a length of transverse colon) was developed by Richie and Skinner (1975) and is now favoured by many bladder surgeons.

Clearly the amount of nitrate and nitrosatable nitrogen in the urine will be independent of the type of operation; the major variable is the bacterial flora, and this was studied by Hill *et al.* (1983). The flora was complex in the urine of patients with ureterocolic anastomosis (Table 8.20), with two or more genera present in large numbers in all of 26 patients and more than four genera in 11 of 26. In contrast there were two or more genera isolated from the urine of eight of 15 ileal loop and zero of 16 colonic loops.

Thus in patients with ureterocolic anastomosis the urine is rich in nitrate, nitrosatable amine and a mixed bacterial population; these are the ideal conditions for the local production of N-nitroso compounds. The first report of cancer in ureterosymoidostomy was by Hammer (1929). A number of further studies of the cancer risk in such patients has been made (Table 8.21); all showed a very large excess of local cancers in contrast to that in the normal population. Stewart *et al.* (1981) have postulated that the N-nitroso compounds known to be produced under such circumstances are responsible for the excess risk of local carcinomas. If this is so, then from the bacteriological study of Hill *et al.* colonic loops should have a much better long-term prognosis than the two alternative diversions studied.

8.7.1.3 Cancer of the cervix

The normal cervix has an acid pH (approximately 4) and harbours a bacterial flora rich in the acid-tolerant lactic acid bacteria — *Lactobacillus* spp and *Streptococcus*

Table 8.20 — The bacterial flora of urine from patients with various urinary diversions

	Type of urinary diversion		
	Ureterocolic anastomosis	Ileal loop	Colon loop
Number of urines examined	26	15	16
Percentage infected	100%	73%	37%
With at least 2 genera	100%	53%	0%
With at least 3–4 genera	88%	27%	0%
With >4 genera	42%	7%	0%

Table 8.21 — Risk of colon cancer in patients with ureterosigmoid anastomosis

Study	Relative risk
Eraklis and Folkman (1978)	180-fold
Parsons *et al.* (1977)	280-fold
Urdaneta *et al.* (1966)	550-fold
Stewart *et al.* (1981)	200-fold

spp. These organisms do not reduce nitrate to nitrite and do not catalyse the N-nitrosation reaction. In disease, in contrast, the pH is close to neutrality and the flora is rich in putrefactive *Bacteroides* spp, *Peptococcus* spp and *Clostridium* spp (Hurley *et al.* 1974, Wilks *et al.* 1984).

Robertson *et al.* (1971) were the first to draw attention to the very high incidence of cervical cancer in Black Africans living in the Transvaal lowveld and associated with the high incidence of *T. vaginalis* infection. Later, Harington *et al.* (1973) and Harington (1975) confirmed, extended and amplified these observations. Harington *et al.* (1973) reported the detection of dimethylnitrosamine in the vaginal vault, and Harington (1975) suggested that the high risk of cervical cancer in women with *T. vaginalis* infection was due to the local formation and activation of N-nitroso compounds by the secondary bacterial changes resulting from the disease.

8.7.2 N-nitroso compounds acting at distant sites

The observations of Caygill *et al.* (1987) of increased risk of cancers at distant sites associated with gastric hypochlorhydria have already been discussed (and also see Table 8.15). If these are due to target organ-specific N-nitroso compounds formed in the gastric milieu, a similar range of sites might be expected to be associated with, for example, bilharzial infection, urine diversion or cervical infection. In fact, none of these patient groups has been studied in this way. The most fruitful patient group in this respect is likely to be the bilharzial patients, because of its size. However, cancer registration is unreliable in the affected regions and so such a study would be a major and lengthy undertaking. The urine-diversion patients are a well defined group, but small in size. Cervical infection is readily treated (with metronidazole) and so is unlikely to yield results.

8.7.2.1 Nitrate and congenital abnormalities

There has been a small number of reports of congenital abnormalities associated with the ingestion of high-nitrate loads by pregnant women, principally tumours of the central nervous system. For example, Scragg *et al.* (1982) reported a localized region in South Australia with an excess of congenital malformations principally of the central nervous system in the new born. They investigated a number of possible causes, and their conclusion was that 'the internal cohesion of the data analysis, and the plausibility conferred by experimental evidence, suggests that underground water, and its elevated concentration of nitrates, may warrant further consideration as a source of human teratogens'. In support of this they reported a 'dose–response' relationship between the nitrate content of the drinking water and the risk of congenital malformations, with a threefold range in risk between the low- and the high-nitrate sources. They noted that in rodents the teratogenic effects of N-nitroso compounds were expressed particularly in the central nervous system (Druckrey *et al.* 1966, Givelber and Dipaolo 1969) whilst Inui *et al.* (1979) showed that nitrate fed to pregnant hamsters caused congenital abnormalities in the offspring. In a follow-up study, Dorsch *et al.* (1984) carried out case–control studies in the region. They confirmed that the nitrate content of drinking water was an important risk factor, with those drinking water with more than 15 mg/l nitrate having a four-fold excess risk.

Preston-Martin and Henderson (1984) carried out case–control studies of patients with intracranial tumours and found an association with nitrite-cured foods or foods thought to be rich in N-nitroso compounds. They discussed their results in the light of previous similar reports from their group and, like Scragg *et al.*, concluded that the possible teratogenic risk or the risk of central nervous system or cranial tumours should be explored further.

In experimental studies there is now a wealth of data showing that N-nitroso compounds given to pregnant rodents result in tumours (usually intracranial) being formed in a high proportion of the offspring. This has been hypothesized as being due to the transplacental passage of the N-nitroso compounds. If this can occur with exogenous N-nitroso compounds then presumably it can occur with the same facility with the endogenously produced analogues. There would be ample opportunity for the endogenous formation of N-nitroso compounds in pregnant women; for example, urinary tract infections are known to be very common in this patient group. For these reasons the reports from Australia (Scragg *et al.* 1982, Dorsch *et al.* 1984) of congenital abnormalities in the offspring of women exposed to high nitrate intakes should not be ignored.

REFERENCES

Adam, B., Schlag, P., Friedl, P., Preussmann, R. and Eisenbrand, G. (1989). Proline is not useful as a chemical probe. *Gut* **30**, 1068–1073.

Amundsen, P. (1947). Poisoning with sodium nitrate. *Nordisk Medisin.* **62**, 1107–1111.

Aquanno, J., Chan, K. M. and Dietzler, D. (1981). Accidental poisoning of two laboratory technicians with sodium nitrite. *Clin. Chem.* **27**, 1145–1146.

Armijo, R. and Coulson, A. H. (1975). Epidemiology of stomach cancer in Chile: the role of nitrogen fertilisers. *Int. J. Epidemiol.* **4**, 301–310.

Barker, M. H. and O'Hare, J. P. (1928). The use of Salyrgan in edema. *JAMA* **91**, 2060.

Bartholomew, B., Hill, M. J., Hudson, M. J., *et al.* (1980). Gastric bacteria, nitrate, nitrite and nitrosamines in patients with pernicious anemia and in patients treated with cimetidine. IARC Scientific Publication No. 31, p. 595.

Bartholomew, B. and Hill, M. J. (1984). The pharmacology of dietary nitrate and the origin of urinary nitrate. *J. Food Toxicol.* **22**, 789–795.

Bavin, P. M., Darkin, D. W. and Viney, N. J. (1982). Total nitroso compounds in gastric juice. In: *N-nitroso Compounds: Occurrence and Biological Effects*. Eds H. Bartsch, I. O'Neill, M. Castagnero, and M. Okada, IARC Scientific Publication No. 41, pp. 337–344.

Binz, C. and Geringer, P. (1901). Die Reduktion des Natriumnitrats in Tierkorper. *Arch. Internat. Pharmacodynam. Therap.* **9**, 441–450.

Bricker, E. M. (1950). Bladder substitution after pelvic evisceration. *Surg. Clin. North Amer.* **30**, 1511–1521.

Brooks, J. B., Cherry, W. B., Thacker, L. and Alley, C. (1972). Analysis by gas chromatography of amines and nitrosamines produced *in vivo* and *in vitro* by *Proteus mirabilis*. *J. Inf. Dis.* **126**, 143–153.

Bruijns, E. (1982). Treatment of kidney stones with ammonium nitrate. Ph.D. Thesis, University of Nijmegen.

Calmels, S., Ohshima, H., Vincent, P., *et al.* (1985). Screening of microorganisms for nitrosation catalysis at pH 7 and kinetic studies on nitrosamine formation from secondary amines by *E. coli* strains. *Carcinogenesis* **6**, 911–915.

Caygill, C. P. J., Hill, M. J., Kirkham, J. S. and Northfield, T. C. (1986). Mortality from gastric cancer following gastric surgery for peptic ulcer. *Lancet* **i**, 929–931.

Caygill, C. P. J., Hill, M. J., Hall, C. N., Kirkham, J. and Northfield, T. C. (1987). Increased risk of cancer at multiple sites after gastric surgery for peptic ulcer. *Gut* **28**, 924–928.

Chen, K. K., Rose, C. L., and Clowes, G. H. (1934). Comparative values of several antidotes to cyanide poisoning. *Am. J. Med. Sci.* **188**, 767–781.

Chen, V. W., Abu-Elyazeed, R., Zavala, D., Haenszel, W., Kitsanes, V., Rice, J., Cuello, C., Montes, G. and Correa, P. (1990). Risk factors of gastric precancerous lesions in a high-risk Colombian population. II Nitrate and nitrite. *Nutr. Cancer* **13**, 67–72.

Comly, H. H. (1945). Cyanosis in infants caused by nitrates in well water. *J. Am. Med. Assoc.* **129**, 112–116.

Cornblath, M. and Hartman, A. F. (1948). Methaemoglobinemia in young infants. *J. Pediatrics* **33**, 421–425.

Correa, P. (1983). The gastric precancerous process. *Cancer Surveys* **2**, 437–450.

Correa, P., Haenszel, W., Cuello, C., Tannenbaum, S. and Archer, M. (1975). A model for gastric cancer epidemiology. *Lancet* **ii**, 58–60.

Craun, G., Greathouse, D., Gunderson, D. (1981). Methemoglobin levels in young children consuming high nitrate well water in the United States. *Int. J. Epidem.* **10**, 309–317.

Davies, J. M. (1977). Two aspects of the epidemiology of bladder cancer in England and Wales. *Proc. Roy. Soc. Med.* **70**, 411–413.

Day, D. (1981). Histopathology of gastric cancer. In: *Gastric Cancer.* Eds J. Fielding, C. Newman, C. Ford and B. Jones, Pergamon Press, Oxford, pp. 95–110.

Diskalenko, A. P. (1968). Methaemoglobinaemia of water nitrate origin in the Moldavian SSR. *Gig. Sanit.* **33**, 30–34.

Dorsch, M. M., Scragg, R. K., McMichael, A. J., Baghurst, P. A. and Dyer, K. F. (1984). Congenital malformations and maternal drinking water supply in rural South Australia: a case–control study. *Am. J. Epidemiol.* **119**, 473–486.

Doyle, M. P., Herman, J.G. and Dykstra, R. L. (1985). Autocatalytic oxidation of hemoglobin induced by nitrite: activation and chemical inhibition. *J. Free Radicals Biol. Med.* **1**, 145–154.

Druckrey, H., Ivankovic, S. and Preussmann, R. (1966). Teratogenic and carcinogenic effects in the offspring after a single injection of ethylnitrosourea to pregnant rats. *Nature* **210**, 1378.

Ellen, G. and Schuller, P. (1982). No volatile N-nitrosamines detected in blood and urine from patients ingesting daily large amounts of ammonium nitrate. *Food Chem. Toxicol.* **20**, 879–882.

Eraklis, A. J. and Folkman, M. J. (1978). Adenocarcinoma at the site of the uretersigmoidostomies for exstrophy of the bladder. *J. Ped. Surg.* **13**, 730–734.

Ewing, M. C. and Mayon-White, R. M. (1951). Cyanosis in infancy from nitrates in drinking water. *Lancet* **i**. 931–934.

Ferrant, M. (1946). Methaemoglobinemia: two cases in newborn infants caused by nitrates in well-water. *J. Pediatrics* **29**, 582–592.

Filipe, I. (1987). Intestinal metaplasia in the histogenesis of gastric carcinoma. In: *Gastric Carcinogenesis.* Eds P. I. Reed and M. J. Hill, Excerpta Medica, Amsterdam, pp. 19–28.

Fine, D. H., Ross, R., Rounbehler, D. P., Silvergleid, A. and Song, L. (1977). Formation *in vivo* of volatile N-nitrosamines in man after ingestion of cooked bacon and spinach. *Nature* **265**, 753–755.

Fraser, P., Chilvers, C., Beral, V. and Hill, M. J. (1980). Nitrate and human cancer: a review of the evidence. *Int. J. Epidem.* **9**, 3–11.

Fraser, P. and Chilvers, C. (1981). Health aspects of nitrate in drinking water. *Sci. Tot. Env.* **18**, 103–116.

Gamgee, A. (1868). Researches on the blood: on the action of nitrites on the blood. *Phil. Trans. Roy. Soc. (London)* **158**, 589–625.

Givelber, H. M. and Dipaolo, J. A. (1969). Teratogenic effects of N-ethyl-N-nitrosourea in the Syrian hamster. *Cancer Res.* **29**, 1151–1155.

Gracey, M., Culli-Jiy, G. J., Suharjono, M., *et al.* (1977). The stomach in malnutrition. *Arch. Dis. Childhood* **52**, 325–327.

Hammer, E. (1929). Cancer du colon sigmoide dixans aprés implantation des uretéres d'une vessie exstrophiée. *J. d'Urologie* **28**, 260–263.

Harington, J. S. (1975). Epidemiology and aetiology of cancer of the uterine cervix. *S. Afr. Med. J.* **49**, 443–445.

Harington, J. S., Nunn, J. R. and Irwig, L. (1973). Dimethyl-nitrosamine in the human vaginal vault. *Nature (London)* **241**, 49–50.

Hawksworth, G. M. and Hill, M. J. (1971). Bacteria and N-nitrosation of secondary amines. *Brit. J. Cancer* **25**, 520–526.

Hawksworth, G. M. and Hill, M. J. (1974). The *in vitro* formation of N-nitrosamines in the rat bladder and their subsequent absorption. *Brit. J. Cancer* **29**, 353–358.

Hicks, R. M., Walters, C. L., Elsebai, I., El Aasser, A.-B., El Merzebani, M. and Gough, T. A. (1977). Demonstration of nitrosamines in human urine: preliminary observations on a possible etiology for bladder cancer in association with chronic urinary tract infections. *Proc. Roy. Soc. Med.* **70**, 413–416.

Hill, M. J. (1984). Aetiology of gastric cancer. *Clinic in Oncology* **2**, 237–249.

Hill, M. J. (1986). *Microbes and Human Carcinogenesis.* Edward Arnold, London.

Hill, M. J. (1988). Gastric carcinogenesis: luminal factors. In: *Gastric Carcinogenesis.* Eds P. I. Reed and M. J. Hill, Excerpta Medica, Amsterdam, pp. 187–200.

Hill, M. J. (1991). Bacterial N-nitrosation and gastric carcinogenesis in humans. *Ital. J. Gastro.* **23**, 17–23.

Hill, M. J., Hawksworth, G. M. and Tattersall, G. (1973). Bacterial, nitrosamines and cancer of the stomach. *Brit. J. Cancer* **28**, 562–567.

Hill, M. J., Hudson, M. J. and Stewart, M. (1983). The urinary bacterial flora in patients with three types of urinary tract diversion. *J. Med. Microbiol.* **16**, 221–226.

Hill, M. J., Fadden, K., Fernandez, F. and Roberts, A. K. (1986). Biochemical basis for microbial antagonism in the intestine. In: *Natural Antimicrobial Systems.* Eds G. Gould, M. Roberts, A. Charnley, R. Cooper and R. G. Board, Bath University Press, Bath, pp. 29–39.

Hurley, R., Stanley, V., Leask, B. and DeLouvois, J. (1974). Microflora of the vagina during pregnancy in: *The Normal Microbial Flora of Man.* Eds E. Skinner and L. Carr, Academic Press, London, pp. 155–186.

Inui, N., Nishi, Y., Taketomi, M. and Mori, M. (1979). Transplacental action of sodium nitrate on embryonic cells of Syrian golden hamster. *Mutation Res.* **66**, 149–158.

Kapanadze, S. K. (1961). On the problem of establishing permissible concentrations of nitrates in water. *Gig. Sanit.* **26**, 7–11.

Kiese, M. and Weger, N. (1969). Formation of ferrihaemoglobin with aminophenols in the human for treatment of cyanide poisoning. *Eur. J. Pharmacol.* **7**, 97–105.

Kosaka, H., Imaizumi, K., Imai, K. and Tyuma, I. (1979). Stoichiometry of the reaction of oxyhemoglobin with nitrite. *Biochim. Biophys. Acta.* **581**, 184–188.

Kosaka, H. and Uozumi, M. (1986). Inhibition by amines indicates involvement of nitrogen dioxide in autocatalytic oxidation of oxyhemoglobin by nitrite. *Biochim. Biophys. Acta.* **871**, 14–18.

Lauren, P. (1965). The two histological main types of gastric carcinoma: diffuse and so-called intestinal type. *Acta. Path. Microbiol. Scnd.* **64**, 31–49.

Leach, S. (1988). Mechanisms of endogenous N-nitrosation. In: *Nitrosamines: Toxicology and Microbiology*. Ed. M. J. Hill, Ellis Horwood, Chichester, pp. 69–87.

Leach, S., Challis, B., Cook, A. R., Hill, M. and Thompson, M. (1985). Bacterial catalysis of the N-nitrosation of secondary amines. *Trans. Biochem. Soc.* **13**, 380–381.

Leach, S., Thompson, M. and Hill, M. (1987). Bacterially catalysed N-nitrosation in the human stomach. *Carcinogenesis* **8**, 1907–1912.

Lecks, H. I. (1950). Methaemoglobinaemia in infancy. *Am. J. Dis. Children* **79**, 117–123.

Lehtola, J. (1978). Family study of gastric carcinoma; with special reference to histological types. *Scand. J. Gastro.* **13**, supplement 50, 1–54.

Magee, P. M. and Barnes, J. M. (1956). The production of malignant primary hepatic tumours in the rat by feeding dimethylnitrosamine. *Brit. J. Cancer* **10**, 114–122.

Magee, P. N. and Barnes, J. M. (1967). Carcinogenic N-nitroso compounds. *Adv. Cancer Res.* **10**, 163–246.

Massey, R. C. (1988). Analysis of N-nitroso compounds in food and human body fluids. In: *Nitrosamines: Toxicology and Microbiology*. Ed. M. J. Hill, VCH Publishers, Cambridge, pp. 16–47.

Munoz, N. and Asvall, J. (1971). Time trends of intestinal and diffuse types of gastric cancer in Norway. *Int. J. Cancer* **8**, 144–157.

Ohshima, H. and Bartsch, H. (1981). Quantitative estimation of endogenous nitrosation in man by monitoring N-nitrosoproline excreted in the urine. *Cancer Res.* **41**, 3658–3662.

Parsons, C. D., Thomas, M. H. and Garrett, R. A. (1977). Colonic adenocarcinoma: a delayed complication of ureterosigmoidostomy. *J. Urol.* **118**, 31–34.

Petukhov, N. I. and Ivanov, A. V. (1970). Investigation of certain psychophysiological reactions in children with water nitrate methemoglobinemia. *Gig. Sanit.* **35**, 26–28.

Pignatelli, S., Richard, I., Bourgade, M. and Bartsch, H. (1987). An improved method for analysis of total N-nitroso compounds in gastric juice. In: *Ninth International Meeting on N-nitroso Compounds: Relevance to Human Cancer.* Eds H. Bartsch and I. O'Neill, IARC, Lyon, pp. 209–215.

Preston-Martin, S. and Henderson, B. E. (1984). N-nitroso compounds and human intracranial tumours. IARC Scientific Publication No. 57, Lyon, pp. 887–894.

Reed, P. I., Smith, P. L. R., Haines, K., *et al.* (1981). Gastric juice N-nitrosamines in health and gastroduodenal disease. *Lancet* **ii**, 550–552.

Richie, J. P. and Skinner, D. G. (1975). Urinary diversion: the physiological rationale for the non-reluxing colonic conduits. *Br. J. Urol.* **47**, 269–275.

Roberts, A. (1988), Ph.D. Thesis (CNAA).

Robertson, M. A., Harington, J. S. and Bradshaw, E. (1971). The cancer pattern of Africa in Africans of the Transvaal Lowveld. *Brit. J. Cancer* **25**, 385–394.

Rodamski, J. L., Greenwald, D., Hearn, W. L., Block, N. L. and Woods, F. M. (1978). Nitrosamine formation in bladder infections and its role in the etiology of bladder cancer. *J. Urol.* **120**, 48–56.

Roe, H. E. (1933). Methaemoglobinemia following the administration of bismuth subnitrate: report of a fatal case. *JAMA* **101**, 352–354.

Rowland, I. R. (1988). The toxicology of N-nitroso compounds. In: *Nitrosamines: Toxicology and Microbiology*. Ed. M. J. Hill, VCH Publishers, Cambridge, pp. 117–141.

Sander, J. (1968). Nitrosaminsynthese durch Bakterien. *Z. Physiol. Chem.* **349**, 429–432.

Savage, W. E., Hajj, S. M. and Kass, E. H. (1967). Demographic and prognostic characteristics of bacteriuria in pregnancy. *Medicine (Baltimore)* **46**, 385.

Schmidt, H., Stich, W., and Kluge, F. (1949). Zur Klinik der Nitratvergiftung Deutsches Med. *Wochensch.* **31/32**, 961–962.

Schultze, W. and Scheibe E (1948). Eine Massenvegiftung mit Natriumnibrit. *Z.f.d. Ges. Inn. Med.* **3**, 580–589.

Scragg, R. K., Dorsch, M. M., McMichael, A. J. and Baghurst, P. A. (1982). Birth defects and household water supply. *Med. J. Australia* **2**, 577–579.

Shearer, L. A., Goldsmith, J. R., Young, C., Kearns, O. A. and Tamplin, B. R. (1972). Methemoglobin levels in infants in an area with high nitrate water supply. *Am. J. Publ. Health* **62**, 1174–1180.

Shuker, D. E. G. (1988). The chemistry of N-nitrosation. In: *Nitrosamines: Toxicology and Microbiology*. Ed. M. J. Hill, VCH Publishers, Cambridge, pp. 48–66.

Sinclair, T. and Tuxford, A. F. (1971). The incidence of urinary tract infection and asymptomatic bacteriuria in a semi-rural practice. *The Practitioner* **207**, 81–90.

Stahlsberg, H. and Taksdal, S. (1971). Stomach cancer following gastric surgery for benign conditions. *Lancet* **ii**, 1175–1177.

Stewart, M., Hill, M. J., Pugh, R. C. B. and Williams, J. P. (1981). The role of N-nitrosamine in carcinogenesis at the ureterocolic anastomosis. *Br. J. Urol.* **53**, 115–118.

Stiles, M. J. (1911). Epispadias in the female and its surgical treatment. *Surg. Gyn. Obstet.* **13**, 127–140.

Stokvis, B. (1902). Zur Casuistik der Autotoxischen Enterogenen Cyanosen. *Internat. Beitr. Inn. Med.* **1**, 59.

Subbotin, F. N. (1961). Nitrates in potable water and their effect on methemoglobin synthesis. *Gig. Sanit.* **26**, 13–17.

Super, M., Heese, H., MacKenzie, D., Dempster, W., DuPlessis, J. and Ferreira, J. J. (1981). An epidemiological study of well water nitrates in a group of South West African/Namibian infants. *Water Res.* **15**, 1265–1270.

Tannenbaum, S. R., Fett, D., Young, V. R., *et al.* (1978). Nitrite and nitrate are formed by endogenous synthesis in the human intestine. *Science* **200**, 1487–1489.

Thomason, H., Burke, V., and Gracey, M. (1981). Impaired gastric function in experimental malnutrition. *Am. J. Clin. Nutr.* **34**, 1278–1280.

Tricker, A. R., Mostaja, M. H., Spiegelhalder, B. and Preussmann, R. (1989). Urinary excretion of nitrate, nitrite and N-nitroso compounds in Schistosomiasis and bilharzia bladder cancer patients. *Carcinogenesis* **10**, 547–552.

Urdaneta, L. F., Duffell, D., Creevy, C. D. and Aust, J. D. (1966). Late development of primary carcinoma of the colon following ureterosigmoidostomy: report of three cases and literature review. *Ann. Surg.* **164**, 503–513.

Vanzant, F. R., Alvarez, W. C., Eusterman, G. B., Dunn, H. L. and Berkson, J. (1932). The normal range of gastric acidity from youth to old age. *Arch. Int. Med.* **49**, 345–359.

Walters, C. L. (1987). Gastric juice N-nitroso compounds. In: *Gastric Carcinogenesis*. Eds P. I. Reed and M. J. Hill, Excerpta Medica, Amsterdam, pp. 163–174.

Walters, C. L., Hart, R. J., Keefer, L. K. and Newberne, P. (1980). The sequential determination of nitrite, N-nitroso compounds and nitrate and its application. IARC Science Publication No. 31, Lyon, pp. 389–399.

Walton, G. (1951). Survey of the literature relating to infant methaemoglobinemia due to nitrate-contaminated water. *Amer. J. Publ. Hlth.* **41**, 986–996.

Watt, P. C., Patterson, C. C. and Kennedy, T. L. (1984). Late mortality after vagotomy and drainage for duodenal ulcer. *Br. Med. J.* **288**, 1335–1338.

WHO (1984). *Guidelines for Drinking Water Quality, Volume 2, Health Criteria and Other Supporting Information,* Geneva.

WHO (1985). *Health Hazards from Nitrates in Drinking Water.* Report of a WHO Working Party, Copenhagen, 5–9 March 1984.

Wilks, M., Thin, R. N. and Tabaqchali, S. (1984). Quantitative bacteriology of the vaginal flora in genital disease. *J. Med. Microbiol.* **18**, 217–232.

Winton, E. F., Tardiff, R. G. and McCabe, L. F. (1970). Nitrate in drinking water. *J. Am. Water Works Assoc.* **43**, 95–98.

Zaldivar, R. (1970). Geographic pathology of oral, esophageal, gastric and intestinal cancer in Chile. *Z. Krebsforsch* **75**, 1–13.

Zaldivar, R. (1977). Nitrate fertilisers as environmental pollutants: positive correlation between nitrates ($NaNO_3$ and KNO_3) used per unit area and stomach cancer mortality rates. *Experientia* **33**, 264–265.

Zaldivar, R. and Robinson, H. (1973). Epidemiological investigation on stomach cancer mortality in Chileans: association with nitrate fertiliser. *Z. Krebsforsch* **80**, 289–295.

Zaldivar, R. and Wetterstrand, W. H. (1975). Further evidence of a positive correlation between exposure to nitrate fertilisers ($NaNO_3$ and KNO_3) and gastric cancer death rates: nitrites and nitrosamines. *Experientia* **31**, 1354–1355.

Zobell, C. F. and Meyer, K. F. (1932). Factors influencing the reduction of nitrates and nitrites by bacteria in semisolid media. *J. Bact.* **24**, 273–281.

Index

FSC
www.fsc.org
MIX
Papier | Fördert
gute Waldnutzung
FSC® C083411

Kolibri 360 GmbH
Ferdinand-Jühlke-Straße 7
99095 Erfurt, Deutschland
produktsicherheit@kolibri360.de